Advanced Technologies in the Fuel and Energy Complex

IOP Conference Series: Materials Science and Engineering
Volume 837

Moscow, Russia
16 - 17 May 2019

ISBN: 978-1-7138-2693-4
ISSN: 1757-8981

Printed from e-media with permission by:

Curran Associates, Inc.
57 Morehouse Lane
Red Hook, NY 12571

Some format issues inherent in the e-media version may also appear in this print version.

Institute of Physics
Dirac House, Temple Back
Bristol BS1 6BE UK

Phone: 44 1 17 929 7481
Fax: 44 1 17 920 0979

techtracking@iop.org

Additional copies of this publication are available from:

Curran Associates, Inc.
57 Morehouse Lane
Red Hook, NY 12571 USA
Phone: 845-758-0400
Fax: 845-758-2633
Email: curran@proceedings.com
Web: www.proceedings.com

TABLE OF CONTENTS

IOP Conf. Series: Materials Science and Engineering **837** (2020) 011001 doi:10.1088/1757-899X/837/1/011001

International conference "Advanced technologies in the fuel and energy complex"

The International conference "Advanced technologies in the fuel and energy complex" was held at the State University of Management (Moscow, Russia) on 16-17 May 2019.

During the conference, representatives of fuel and energy companies, the scientific community, government agencies, as well as teachers, graduate students and students discussed issues of further development of the fuel and energy complex.

Within the framework of the conference, the following problems were considered:

- Technologies of production and transmission of fuel and energy resources;
- Technological, innovative and digital modernization of fuel and energy complex facilities;
- Prospects for the introduction of advanced technologies in the industry;
- Organization of production.

The program committee summed up the conference and prepared a resolution of the conference, which made the following conclusions about the state of the fuel and energy complex and recommendations for its development:

- Lack of necessary technologies, which are an integral process of stimulating the development of the fuel and energy complex;
- Decrease in the efficiency of industrial production due to obsolescence and depreciation of equipment;
- The need for modernization of power plants;
- The development of new methods for generating electric energy and the use of fuel and energy resources;
- Ensuring the localization of production capacities in the territory of national states;
- The creation of new transport systems that ensure the transfer and distribution of resources on a national and global scale;
- Creation of an environmental preservation system for future generations;
- Search for the most effective technologies for the use of oil, gas, coal and other natural resources;
- Ensuring the transition to innovative and digital technologies;
- Development of a system of interaction with various national states to achieve stability, reliability and security of the fuel and energy complex.

Thus, the conference allowed formulating scientific recommendations for fuel and energy companies, state and regional authorities, international and supranational organizations, the scientific and professional community.

IOP Conf. Series: Materials Science and Engineering **837** (2020) 011002 doi:10.1088/1757-899X/837/1/011002

Peer review statement

All papers published in this volume of *IOP Conference Series: Materials Science and Engineering* have been peer reviewed through processes administered by the proceedings Editors. Reviews were conducted by expert referees to the professional and scientific standards expected of a proceedings journal published by IOP Publishing.

IOP Conf. Series: Materials Science and Engineering **837** (2020) 012001 doi:10.1088/1757-899X/837/1/012001

Applying the potentiating procedure for optimal management of power consumption of technocenose

V I Gnatyuk[1], S A. Polevoy[2], O R Kivchun[3] and D V Lutsenko[4]

[1] Kaliningrad State Technical University, 1, Sovetsky Prospekt, Kaliningrad, 236022, Russia
[2] Financial University under the Government of the Russian Federation, 15, Verkhnyaya Maslovka Verkhnyaya, Moscow, 127083, Russia
[3] Immanuel Kant Baltic Federal University, 14, A. Nevskogo street, Kaliningrad, 236016, Russia
[4] Kaliningrad Innovation Center «Technocenose», 8, Zemelnaya street, Kaliningrad, 236029, Russia

E-mail: okivchun@gmail.com

Abstract. The concept of energy saving potential is introduced and justified in the article. A technique for assessing the energy saving potential of technocenoses is presented. The basis of this technique is the following stages: analysis of information on power consumption, rationing and analysis of power consumption. As an experimental verification of the methodology, the results of determining energy saving potentials for the regional electrical complex of Gasprom extraction Urengoi are presented.

1. Introduction

The development of digital technologies in the end of 20th century contributed to the technical progress rising to the new level. Due to that, industry and economics are developing, new scientific theories and technologies are being elaborated, innovative means of communication are upgrading, medicine is improving, new branches are appearing.

Furthermore, the development of digital technologies has improved dramatically the efficiency of resource management in big technical systems – technocenose. Technocenose is a limited in space and time interconnected complex consisted of further indivisible technical items united by weak connections (region, city, municipality, organization, enterprise, company, trading network etc.).

However, comprehensive analysis [1-7] has showed that use of digital technologies alone does not solve the whole spectra of problems of resource management in technocenose. Taking this into the account, one of the solutions of these problems is the development of scientifically proved methodology of management of electrical energy consumption in technocenose based on the potential of electrical energy saving

Therefore, the aim of the research in this area is the development of methodological basics of resource management in technocenose based on the potential of electrical energy saving. The object of the research is considered the system of resource management in technocenose; the subject is the methodology of optimal management of electrical energy consumption in technocenose.

2. Materials and methods

The aim of the work is to analyze the potentiation procedure for optimal control of the energy consumption of the technocenosis. The objectives of the study are as follows:

- Analysis of technocenosis approach to determining the potential for energy conservation;
- Studying the algorithm for the implementation of potentiation procedures;
- Determine the energy saving potential for the regional electrical complex of Gazprom extraction Urengoy.

The study is based on the use of scientific methods and information sources, which allowed the authors to formulate recommendations and suggestions in the field of industrial complex development.

3. Results

The methodology of optimal power management of technocenoses consists of the following calculation blocks: data preparation, identification of objects that abnormally consume electricity, forecasting and rationing [2-3]. To improve the quality of managing the process of power consumption of technocenosis objects, the "potentiation" procedure has been developed and is proposed as an addition to the above method.

Potentiation is a procedure that calculates the required integral value of power consumption, the value of which can reduce the power consumption of all objects [2]. The determining concept for potentiation is the energy-saving potential, which is the absolute difference between the values of power consumption and the lower boundary of the variable confidence interval (figure 1) [2].

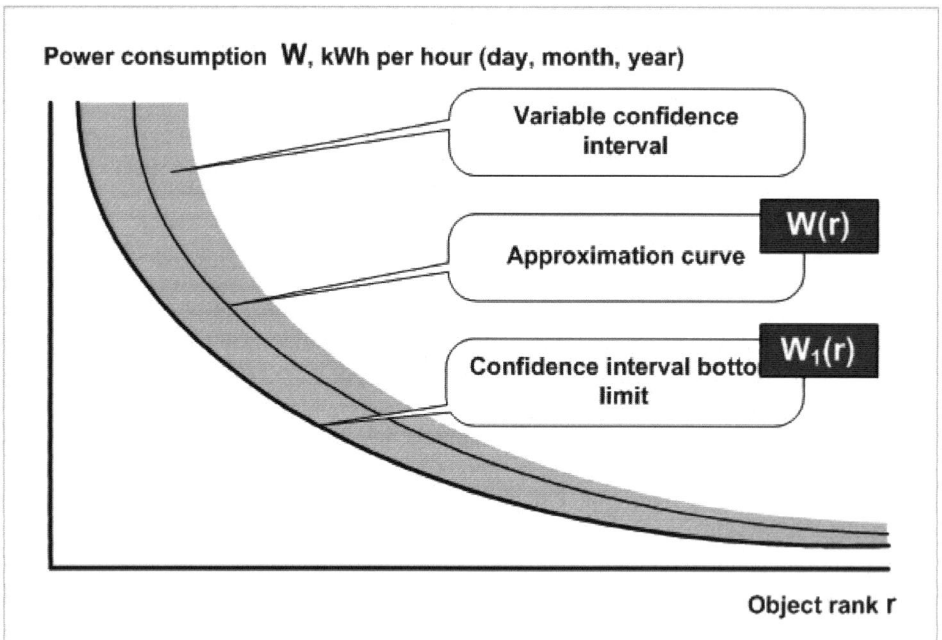

Figure 1. Rank curves for calculating energy conservation potential.

The mathematical calculation of the power consumption potential includes the calculation of the integral from zero to infinity under the rank curve, which can be obtained from empirical values of power consumption or correspond to the lower boundary of the variable confidence interval.

A great deal of projects and scientific works of foreign scientists and businessmen are devoted to the problems of calculation of the potential for energy savings. A group of foreign specialists in the

IOP Conf. Series: Materials Science and Engineering **837** (2020) 012001 doi:10.1088/1757-899X/837/1/012001

field of increase of effectiveness of consumption of natural resources (Weizsecker E.U., BlochLevins A., Mountain R., HunterLovins L.) introduced into everyday life so called «Factor four» that implied to live twice as good and spend half as much resources at the same time. Actually, this is a new approach to calculate the potential for energy savings putting at the top the increase of efficiency of resources by means of constructive and organizational actions. Nowadays this particular method is used in the countries below: USA, Germany, Great Britain, Netherlands, Austria, Denmark, Finland [1; 8-11].

The invaluable contribution in the development of the theory of calculation of the potential for energy savings was made by scientists in huge Research Centers in Europe and USA. As a part of the famous National laboratory of the ministry of energy of USA (Lawrence Berkeley National Laboratory – LBNL) Arthur Rosenfeld arranged a department devoted to energy savings in constructions which became one of trend-setter in this field. That was LBNL that worked in close rapport with European specialists and developed the technology of industrial production of magnetron application of heat-reflective coatings on large-sized flat glasses that became a significant breakthrough in the window field [1; 8-11; 13-15].

Scientific papers of F. Haiek, M. Fridman, R. Lukas offer approaches to calculate the potential for energy savings on the basis of company-internal management of expenses of production, technological and economical efficiency due to the introduction of progressive energy technologies.

The representatives of Keynesian and Neo-Keynesian economics (D. Mid, D. Tobin, R. Solow, N. Kaldor, J.M. Keynes) believe that the state is able to undertake functions of invention of the conception of calculation of the potential for energy savings for all participants of economical activity.

In Russia current approaches to calculate the potential for energy savings can be divided into two large groups according to the kind of approach: classical or technocenose one.

The classical approaches (scientific works of N.I. Danilov, Ya.M. Shelokov, V.V. Dobrodei, O.L. Danilov, A.B. Garyaev, I.V. Yakovlek, A.N. Dmitriev, V.V. Litvak, V.S. Stepanov and others), that are based on the Gaussian mathematical statistics, suggest normal distribution of parameters of a real object and conversion of its power consumption on the basis of some model analogue in the model and basic technological process. The calculation of the potential for energy savings of objects, that have the same structure but work in different conditions, lead to different assessments so it is necessary to take into account specific feature of each one. Received assessments of the potential for energy savings for a complex of objects are individual and their sum doesn't let objectively estimate this amount for a company overall.

The technocenose methods (scientific works of B.I. Kudrin, V.I. Gnatuk, V.V. Fufaev, O.E. Lagunkin, M.G. Oshurkov, B.V. Jilin, A.E. Severin, S.N. Grinkevich and others) are based on the technocenose approaches, the digital mathematical statistics and the theory of hyperbolic endlessly divisible N-assignments. It is assumed that the complex of objects of a concerned enterprise forms a system of particular kind called technosenosis that appears to be limited in time and space integrated complex of functionally completed technic products that are poorly connected. The connections in a technocenose are specific depending on the technological self-sufficiency and diversity of solving tasks [8-11].

The main feature of the technocenological approach to determining the energy saving potential is represented by the calculation method. It is calculated as the sum of the separately taken differences of the actual power consumption and the values based on the optimal indicators of energy conservation.

The energy saving potential, which is calculated for the objects of the technocenosis, is designated - "Z-potential" - by the first letter of the scientist's name Zipf [2] and is calculated as:

$$\Delta W_l = \int_0^\infty W(r)dr - \int_0^\infty W_l(r)dr, \qquad (1)$$

Whereas W_l - is technocenose electricity saving potential;

W(r) - is approximation curve obtained for empirical values of facilities' electricity consumption (r – rank);

$W_1(r)$ - is the limit of variable confidence interval obtained based on database processing [6; 10; 11].

Based on the analysis of the multiple implementation of standard procedures for the optimal power management methodology to improve the quality of the power management process, refinement procedures have been developed (figure 2) [3]. Figure 2 shows the structural diagram of the correspondence of standard and subtle rank analysis procedures.

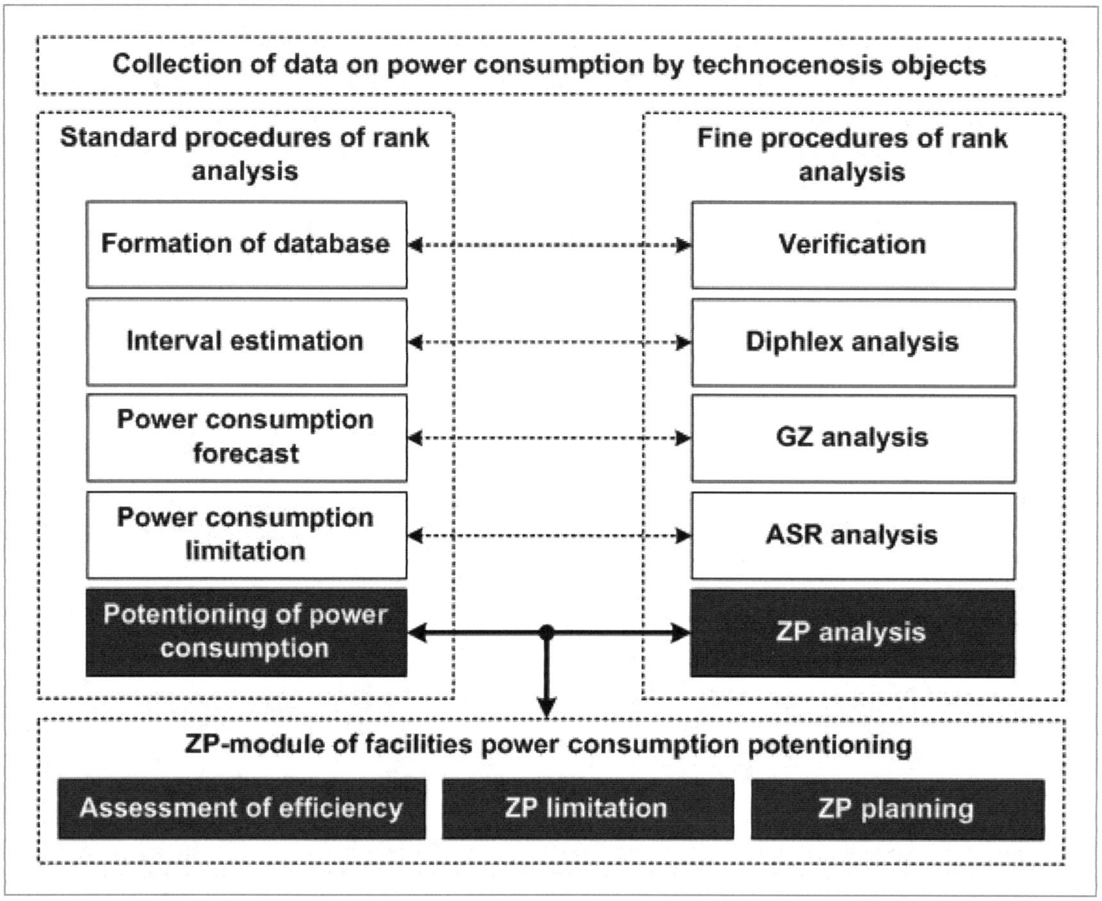

Figure 2. Conformity of the procedures for evaluating the energy saving potential.

The study of the energy saving potential has allowed us to develop a refinement procedure - ZP analysis. This analysis is carried out at the stage of potentiation of power consumption to draw up an energy saving plan (ZP-plan). Its basis is a methodology for assessing potential, which includes two levels [2] (figure 3).

On the first, the potential Z1 is calculated, and on the second, the potential Z2. The first potential is the lower limit of the variable confidence interval, built on the actual values of power consumption. The second potential is the boundary of the variable confidence interval obtained as a result of normalization of power consumption (ZP-normalization) [2].

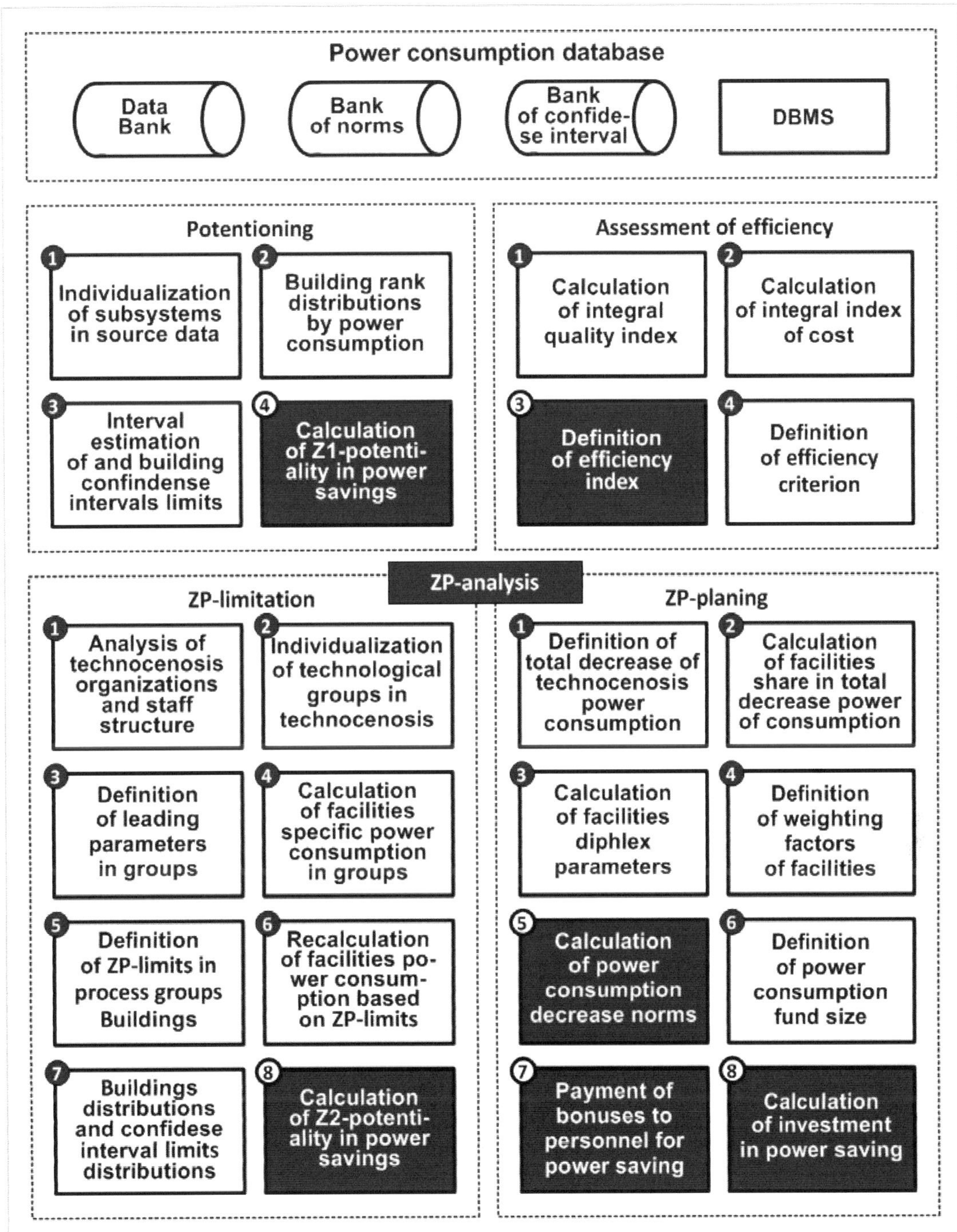

Figure 3. Potential based power analysis block diagram.

At the last stage of the ZP analysis - the preparation of the ZP plan, two stages are implemented. First, the power consumption values of technocenosis objects should be reduced to the level of the Z1 potential, and then to the level of the Z2 potential. Moreover, the reduction is carried out individually for each object [2].

The input data for the ZP-module is generated from the database on power consumption and may include information on the organizational and staff structure, geo-information, information about leading parameters, power consumption values, etc.

The first stage of the ZP-module is the implementation of the potentiation procedure, in the framework of which the initial data on power consumption are imported, the construction of rank distributions and obtaining the Z1-potential.

At the second stage, the effectiveness of the process of managing the power consumption of the technocenosis is assessed based on a criterion that shows the ratio of integrated quality and cost indicators [2].

In the third stage, ZP analysis is implemented. Moreover, ZP-rationing is carried out first, and then ZP-planning. It is important to note that ZP-rationing has the goal of determining the Z2-potential and is a preparation stage for ZP-planning. ZP-rationing operations are based on the study of the organizational system of technocenosis. The result of which are formed technology groups. For each group, a leading parameter is determined. According to which the specific power consumption is calculated [2].

At the final stage of the ZP analysis, which is also the output of the ZP module, ZP planning is carried out. Important information for ZP planning are the obtained values of Z1 and Z2 potentials. As a result, on the basis of the chosen strategy, the volume of reduction in energy consumption is calculated. Next, the calculation of the difflex parameters is carried out, according to which the weight coefficients of the norms for reducing power consumption are determined. The above operations make it possible to calculate the amount of the energy conservation fund, which can be used to reward staff or be used to modernize the technocenosis electrical equipment [2].

Thus, the obtained ZP-plan is the result of the potentiation procedure and includes individual norms for reducing energy consumption and the amount of bonus staff.

4. Discussion

Quantitative estimate of authenticity of results of the offered methodology was realized on the data of energy consumptions of the local electro technical complex Gasprom extraction Urengoi and it included an examination of adequacy of the methodology of the ZP-analysis [2; 12].

The examination of adequacy is based on the analysis of statistic hypothesizes that prove legitimacy of use of chosen probability laws. Assessments of working efficiency are based on the comparative analysis of quality in the long run of 5 years of 2 options of the ZP-plan (table 1) with the use of the system of indexes below.

Table 1. Indicators of the first and second potentials.

Index	A strategy to reach the level of Z1-potential	A strategy to reach the level of Z2-potential
Δ, %	99.97	99.95
$\overline{S}$, %	0.28	0.21
S^{max}, %	5.92	6.41
n_{ZP}, %	85.95	90.08

The proportion of the realized Z-potential for energy savings is Δ, average $\overline{S}$ and maximum S^{max} relative deviation of energy consumption of objects from the bounds of the Z-potential, the proportion of objects that reached the bound of the Z-potential n_{ZP} [2; 12].

The accounts of statistic criteria haven't exceeded critical ones during the examination of adequacy. The amounts of quality indexes of both variations of the ZP-plan conform to each other and that complies with the concept of the methodology of the ZP-analysis and determines by gist and specific features of the Z1-

IOP Conf. Series: Materials Science and Engineering **837** (2020) 012001 doi:10.1088/1757-899X/837/1/012001

and Z2-potentials. The received results let make a conclusion that the developed methodology is adequate and its use is explained for the control for energy consumption of a large industrial enterprise.

The economical assessment of the results of the methodology involves quantitative comparison between expenses and the amount of incoming profit. This comparison is made due to the pay-off period of enterprises on the objects of the local electro technical complex Gasprom extraction Urengoi with consideration to discounting of receiving incomes, the amount of the index of profitableness of investments and other indexes (tables 2, 3). Direct repayment of investments in energy savings was expected to be according to calculations [2; 12].

Table 2. Indicators of the first and second potentials.

Index	Steps of the ZP-plan					
	0	1	2	3	4	5
Energy saving Fund, millions of rubles	0	6.87	6.85	6.99	4.3	0.43
Discounted income, millions of rubles	-4.15	2.68	2.19	1.8	0.91	0.08
Net profit, millions of rubles	-4.15	-1.44	0.76	2.54	3.45	3.59
Amount of income in Energy Saving Trust	0	0.47	0.46	0.44	0.42	0.41
Profit after energy saving	0	6.9	6.82	7.0	4.5	0.41

Table 3. Indicators for obtaining the second potential.

Index	Steps of the ZP-plan					
	0	1	2	3	4	5
Energy saving Fund, millions of rubles	0	49.13	49.31	49.66	50.44	45.13
Discounted income, millions of rubles	-34.39	29.15	23.01	17.47	12.13	5.93
Net profit, millions of rubles	-34.39	-5.24	17.78	35.24	47.37	53.8
Amount of income in Energy Saving Trust	0	0.7	0.65	0.58	0.47	0.3

Analysis of tables 1-3 showed that when investing funds in 4.15 million rubles. upon receipt of the first potential and 34.39 million rubles. - for the second, the benefit will amount to 3.59 and 53.8 million rubles. In addition, these results show that the second potential obtained has the best indicators, as it allows for the modernization of the electrical equipment of objects, and also makes it possible to rationally use the profit over the entire planning interval. At about the second step of the plan, obtained with the help of the second potential, the first investments will be paid off and new income will appear. The conducted experiment confirms the economic feasibility of applying the energy saving potential for optimal management of the regional electrical engineering complex Gazprom dobycha Urengoy [2; 12].

5. Conclusion

Thus, in order to improve the quality of optimal power management of technocenoses, a potentiation procedure was developed and implemented. Its subtle addition is the ZP analysis, which is based on the methodology for assessing the energy saving potential. The main result of the ZP analysis is the individual norms for reducing power consumption for technocenosis facilities and the amount of bonus payments to personnel.

IOP Conf. Series: Materials Science and Engineering **837** (2020) 012001 doi:10.1088/1757-899X/837/1/012001

References

[1] Gaines B R and Kohout L J 1977 Fuzzy decade: a bibliography of fuzzy systems and closely related topics *International Journal of Man-Machine Studies* **9** 1-68

[2] Gnatiuk V I 2012 *Technocenose energy saving potential Treatise* (Kaliningrad: CPC Technocenosis) 56

[3] Gnatiuk V I and Lutsenko D V 2010 *Forecasting electricity consumption based on GZ analysis* (Moskva: Rossijskaya akademiya nauk) 144

[4] Gnatyuk V I, Lutsenko D V, Kivchun O R, Primak L V, Lopatin A E and Fraltsov A G 2014 Ranked parametric distribution technocenosis on bi-furkatsionnom stage of development *Mekhanizaciya stroitel'stva* **10** 44-7

[5] Gnatyuk V I, Lutsenko D V, Kivchun O R and Vasiliev V N 2015 Methods of the monitoring electricity electrotechnical complex of the Kaliningrad region *Promyshlennaya energetika* **3** 26-35

[6] Kivchun O R and Gnatyuk V I 2017 Intellectual technologies of monitoring energy consumption of electrical objects portside complex *Morskie intellektual'nye tekhnologii* **3(37)** 1 138-5

[7] Kivchun O R, Gnatyuk V I, Kretinin G V and Lutsenko D V 2018 Potential of energy saving as a tool for increasing *International journal of energy economics and policy. Mersin: Cag University* **8(1)** 137-43

[8] Kosharnaya Y V 2015 Develop a system of rationing energy consumption indicators and evaluation of energy saving volumes on the example of metallurgical enterprise *Promyshlennaya energetika* **8** 13-22

[9] Kosharnaya Y V 2016 Optimization of structure of energy consumption of metallurgical enterprises to evaluate the energy saving potential *Promyshlennaya energetika* **10** 22-9

[10] Kudrin B I 1993 *Introduction to technetics* (Tomsk: TSU Publishing House) 552

[11] Kudrin B I 2016 Electricity today and problemmy electrical supply *Promyshlennaya energetika* **10** 5-9

[12] Lutsenko D V, Kivchun O R and Primak L V 2014 Potenshirovanie in power consumption management method technocenosis *Mekhanizaciya stroitel'stva* **8** 19-27

[13] Morkovkin D E, Gibadullin A A, Romanova Ju A, Erygin Yu V and Ziadullaev Ul S 2019 Formation of a national environmental strategy for the fuel and energy complex *IOP Conference Series: Materials Science and Engineering* **537** 042064

[14] Lopatkin D S, Shushunova T N, Shaldina G E, Gibadullin A A and Smirnova I L 2019 Renewable and small energy development management *Journal of Physics: Conference Series* **1399** 033061

[15] Morkovkin D E, Kerimova Ch V, Dontsova O I and Gibadullin A A 2019 The formation of factors affecting the sustainable development of the generating complex of the electric power industry *Journal of Physics: Conference Series* **1399** 033042

IOP Conf. Series: Materials Science and Engineering **837** (2020) 012002 doi:10.1088/1757-899X/837/1/012002

Features of the implementation of environmental policies by Russian fuel and energy enterprises

Zarina Tsopanova and Nataliya Kharitonova

Financial University under the Government of the Russian Federation, 49, Leningradsky Prospekt, Moscow, 125993, Russia

E-mail: ztsopanova@inbox.ru

Abstract. The study discovers the problem of combating greenhouse gas emissions, which is most relevant for companies in the fuel and energy complex. The country and industry structure of carbon dioxide emissions was analyzed based on available statistical data. Scenario forecasts of changes in the level of emissions developed by the international energy agency are presented here, possible fiscal measures aimed at preventing further development of the "greenhouse effect are also considered. It was noted that Russian fuel and energy companies need to switch to regular accounting and reporting of actual greenhouse gas emissions, which can ensure that they maintain their achieved level of competitiveness in the world energy markets.

1. Introduction

Environmental pollution has now ceased to be a purely environmental problem. Resisting climate change under the influence of anthropogenic activities is now part of the business management process, which fully applies to the fuel and energy enterprises (hereinafter, the fuel and energy complex). Escalation of environmental liability is caused not only by the good faith of management of the respective companies, but also the legislation's development in this field. More and more countries are developing and imbedding law's drafts that limit the permissible level of environmental pollution, as well as the maximum allowable amounts of greenhouse gas (GHG) emissions, in particular, carbon dioxide CO [1-2].

The implementation of such measures for Russian fuel and energy companies will cause such negative effect as growth in current and investment costs because of the need to switch to technologies that exclude negative impact on the environment beyond the established standards, which inevitably leads to a decrease in the efficiency of management, as well as their level of competitiveness in comparison with enterprises operating in countries that do not impose such strict limitations [3].

2. Materials and methods

According to the Glossary of the Intergovernmental Panel of Climate Change (IPCC), greenhouse gases (GHGs) are gaseous components of atmospheric air of both natural and anthropogenic origin which absorb and emit infrared radiation (heat The Earth's surface). The main GHGs in the Earth's atmosphere are water vapor (H_2O), carbon dioxide (CO_2), nitrous oxide (N_2O), methane (CH_4) and ozone (O_3) [2]. The largest share of GHG emissions is carbon dioxide, CO_2, and is about 66% [2; 4-6].

In 2018, the global level of CO_2 emissions increased by 58% compared to 1990 and amounted to 33.3 GT, according to the IEA assessment. The average growth in the period 1990-2018 was 1.9%.

according to preliminary estimates of IEA analysts, in 2019, the volume of CO_2 emissions remained at the level of 2018 (33 GT) [3].

At the country level, GHG emissions were distributed as follows:

- China and the United States were responsible for more than 40% of global CO_2 emissions. At the same time, China is the leader in terms of CO_2 emissions in absolute terms (in 2018 - 11.3 Gt). U.S. emissions in 2018 were 5.3 Gt;
- The third position is held by India - 2.6 Gt (2018);
- Russia is in 4th place in terms of GHG emissions (2018) [2; 10].

However, when analyzing per capita emissions, the leader among the countries under consideration is Saudi Arabia, followed by Australia, the United States and Canada. When calculating CO_2 emissions per capita, Russia ranks 6-th among the countries under consideration.

The results of industry analysis show that the energy sector is the main source of GHG emissions into the atmosphere (37% of total emissions), ahead of transport and manufacturing in the aggregate of its constituent industries [2; 10].

At the local level, the problem of energy companies emissions is particularly relevant in the regions where coal resources are used. A possible solution to reduce the negative impact on nature can be more active use of wind and solar energy, namely, the creation of individual devices consisting of wind and solar stations in order to provide electricity to individual consumers.

3. Results

The international energy agency (IEA) is conservative in its forecasts: the reduction of CO_2 emissions in the energy sector by 2040 is possible exclusively within the framework of the implementation of the sustainable development scenario which assumes the implementation of the goals of the Paris agreement, which determined that CO_2 emissions in the energy sector by 2040 should be reduced by more than half compared to the level of 2018. If the current environmental protection policy is maintained, it is assumed that CO_2 emissions in the energy sector will increase by an average of 100 million CO_2 in the period from 2018 to 2040, which is less than the average growth rate observed since 2010 (350 million tons of CO_2), but this is not enough to achieve the goals of the Paris agreement and the overall level of CO_2 emissions will increase from 33 GT in 2018 to 36 GT in 2040 [2; 7; 10].

As for Russia, CO_2 emissions (excluding changes in land use) in Russia by 2010 decreased by 34.8% compared to 1990, but since 2010 there has been a rising of the quantity of greenhouse gas entering the atmosphere (the average annual rate is 0.29%).

It should be emphasized that the share of CO_2 emissions from the Russian energy sector is higher than the global average, at 46% versus 37%. In 2019, Russia adopted the Paris agreement and set some targets that set the direction for improving the state of the environment. In particular, by 2030, Russia plans to reduce greenhouse gas emissions by 25-30% relative to the 1990 baseline. However, the specific target for emissions by 2030 (18.5 t of CO_2/person) is several times higher than the value needed to keep the temperature increase within 2°C (4.7 t of CO_2/person in 2030) [2;7-8; 10].

In Russia, there is no mandatory reporting of GHG emissions, so trends within specific energy companies can be judged by one-time publications that are not of a regular nature. Compliance with the terms of the Paris agreement, as well as compliance with tougher environmental legislation will require Russian energy companies a more responsible attitude to the organization of the industrial ecological supervision.

4. Discussion

In the conditions of the global decarburization policy's realization, the energy sector is at the greatest risk of losing its previously achieved level of competitiveness, which needs to introduce more modern technologies, reduce the consumption of fossil fuels, and take an active part in the realization of the state environmental program.

At the moment, the main method of government incentives to reduce GHG emissions, which is common in the world, is carbon pricing, which distinguishes the following two mechanisms: emissions trading and carbon tax. When using emission quotas, a fixed level of target emissions is set, this may gradually decrease over time. Companies receive quotas for GHG emissions through auctions or free of charge with the possibility of further resale. If a carbon tax is applied, companies contribute a set amount to the budget for each ton of CO_2 released into the atmosphere. Thus, the "carbon price" is set directly through the tax rate and the amount of carbon dioxide produced by burning fossil fuels. In 2019, by the world Bank's ratings that the number of carbon pricing initiatives introduced worldwide is approaching 60 positions [8-9].

In addition to carbon pricing mechanisms, it is possible to reduce emissions through legal prohibitions and restrictions. The most significant initiative limiting global GHG emissions was the Paris agreement, which regulates the goals for reducing the content of carbon dioxide in the atmosphere. In the end of 2019, the agreement was ratified in 189 countries. UN analysts predict that in order to achieve the goals of the Paris agreement, it is necessary - reduce GHG by 25% for 2019-2050 (from 55.3 to 41 GT of CO_2 equivalent) to keep the average annual temperature increase within 2°C. Keeping the temperature increase at 1.5°C will require reducing GHG emissions by more than half [2; 10-11].

In Russia now, there is no fee for greenhouse gas emissions, and disclosure of data and accounting for emissions is Advisory in nature. The main reason for this situation is that, on the one hand, the introduction of a fee in the form of an emissions tax would help to achieve the goals set under the Paris agreement, which Russia signed in 2019, and on the other hand, such a fee would increase the fiscal burden not only on energy companies, but also on the final consumers of fossil fuels. The refusal to introduce an additional tax will preserve the financial stability of these companies and end-consumers, but will entail external risks associated with the aggravation of foreign economic relations and a decrease in trade volumes, in particular, with the EU countries.

Modern technologies for reducing emissions are divided into two complementary groups:

- Conservation of gross greenhouse gas emissions (switching from coal to gas and renewable energy sources, electrification of transport). In Russia, such technologies as rational use of APG, APG injection into the oil and gas reservoir, energy efficiency improvement, and CO_2 storage in the oil and gas reservoir are used. In the coal industry, cogeneration, installation of filters at thermal power plants, and utilization of coal mine methane have become widespread. Less popular is the use of renewable energy sources (about 1% in the Russian energy sector);
- Sequestration technologies: use of natural sinks (reforestation) and carbon capture and storage technologies to reduce the carbon content in the atmosphere (CCS, "carbon traps").

Thus, in order to reduce the negative impact of energy facilities on the environment, it is advisable to use renewable energy sources, the scheme of which is presented in the figure 1.

5. Conclusion

Thus, these data confirm the initial thesis that the problems of air pollution have gone beyond purely environmental problems and can have a significant impact on the activities of a single fuel and energy company, as well as on the economies of individual countries and the world as a whole.

Currently, the market for professional investments in the development of environmentally friendly technologies and the implementation of socially responsible policies in Europe is $14 trillion, in the United States - $12 trillion, and the return on shares of companies that meet the criteria for sustainable development often exceeds the return on their less socially responsible counterparts [10-11].

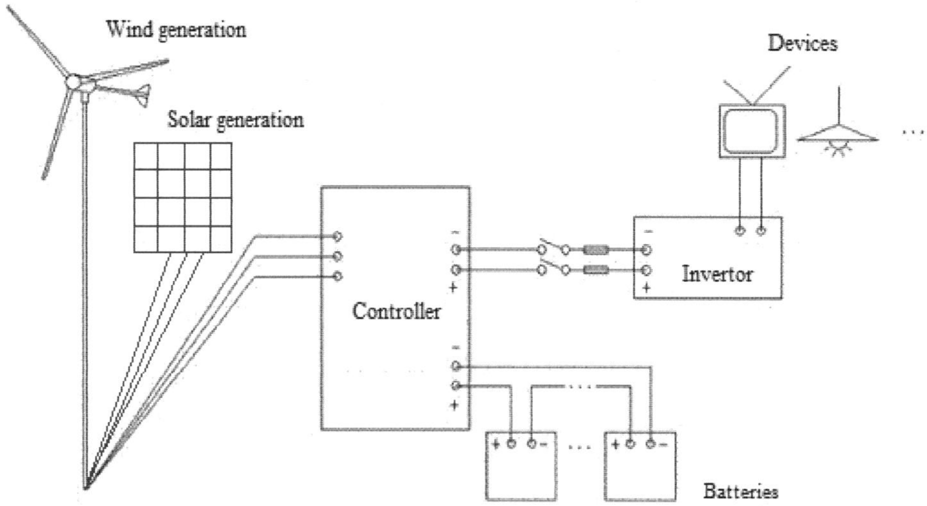

Figure 1. Scheme of operation of an individual power generating device operating on renewable energy sources.

In this regard, Russian fuel and energy companies will have to adapt to changing conditions in order not to lose their positions in the global energy markets, which implies regular accounting of emissions, diversification of strategies and business plans in accordance with the need to switch to "green" technologies for the extraction, processing and consumption of fossil fuels.

References

[1] Gibadullin A A et al 2020 *IOP Conf. Ser.: Earth Environ. Sci.* **421** 032051
[2] *Russian statistical yearbook* 2018 (Moscow: Rosstat) 694
[3] Lopatkin D S *et al* 2019 *J. Phys.: Conf. Ser.* **1399** 033061
[4] Sadriddinov M I *et al* 2020 *IOP Conf. Ser.: Mater. Sci. Eng.* **734** 012051
[5] Ivanova I A *et al* 2020 *IOP Conf. Ser.: Earth Environ. Sci.* **421** 032039
[6] *Energy Strategy of Russia until 2035* Retrieved from: https://minenergo.gov.ru/node/1920
[7] Romanova Ir N *et al* 2020 *IOP Conf. Ser.: Mater. Sci. Eng.* **734** 012166
[8] Gibadullin A and Pulyaeva V 2019 Obstacles to the formation of a common electricity market of the Eurasian Economic Union *E3S Web of Conferences* **114** 02002
[9] Ivanova I A *et al* 2019 *J. Phys.: Conf. Ser.* **1399** 033038
[10] *Ministry of Energy of the Russian Federation* Retrieved from: https://minenergo.gov.ru
[11] Gibadullin A A *et al* 2020 *IOP Conf. Ser.: Mater. Sci. Eng.* **734** 012170

IOP Conf. Series: Materials Science and Engineering **837** (2020) 012003 doi:10.1088/1757-899X/837/1/012003

Sulfur-containing emulsions under alternative non-thermal initiation: a way to increase the energy efficiency of chemical processes

Alexey Zanin[1], Dmitry Lopatkin[1], Pavel Sobolev[1], Manuchehr Sadriddinov[2], Arthur Gibadullin[3] and Dmitry Morkovkin[4]

[1] Dmitry Mendeleev University of Chemical Technology of Russia, 9, Miusskaya Square, Moscow, 125047, Russian Federation
[2] Tajik State Finance and Economics University, 64/14, Nakhimova Street, Dushanbe, 734067, Republic of Tajikistan
[3] State University of Management, 99, Ryazan Avenue, Moscow, 109542, Russian Federation
[4] Financial University under the Government of the Russian Federation, 49, Leningradsky avenue, Moscow, 125993, Russian Federation

E-mail: dreiven118@mail.ru

Abstract. Currently, there is a tendency to develop chemical and chemical-technological processes that consume as little energy as possible. Recently, ionizing and microwave radiation began to be used for these purposes in the chemical industry, since conventional heating can be slow and inefficient, which increases energy consumption. Polymer sulfur, which is one of the modifications of elemental sulfur, has a number of valuable properties, however, the existing methods for its synthesis are characterized by high energy consumption and therefore are disadvantageous. In order to develop a more efficient method for the synthesis of polymer sulfur, the processes that occur in sulfur-containing emulsions under the influence of ionizing and microwave radiation, and the products formed in these processes are investigated.

1. Introduction

Sulfur-containing polymers are widely known for many years [1]. Polymer sulfur has high impact strength, high resistance to aggressive media, lack of thermal shrinkage in composites, insolubility in rubber, etc., so it can be used in industry for the vulcanization of rubber and as a plasticizer in concrete or ceramics. But high energy consumption of polymer sulfur synthesis processes limits its application.

Less energy intensive is the radiation initiation of chemical processes. Radiolysis and subsequent chemical transformations have been studied very well [2]. In the case of radiation initiation, in principle, an increase in the temperature of the process is not required, while it is possible to carry out the process at high speeds at low temperatures, and the likelihood of side processes is reduced. The speed of the process does not change over time, and the reaction can be completely stopped by removing the reaction system from the irradiation zone or removing the source. This makes it possible to relatively quickly and easily control the initiation rate and properties of the resulting polymers [3-8].

Polymerization processes under the influence of microwave radiation were considered in common earlier [9]. Microwave initiation is widely used in polymer synthesis [10–14], it is applicable for liquid, solid, and paste media. Microwave heating is fast and efficient. The use of microwave energy instead of traditional methods can significantly simplify the technological scheme, eliminating all processes and devices associated with the preparation of the coolant, as well as significantly reduce the amount of harmful wastes [14].

2. Materials and methods

In the studies elemental sulfur S_8, benzene C_6H_6, toluene $C_6H_5CH_3$, and a number of the emulsifiers were used. Benzene and toluene was chosen as the reaction media for carrying out the processes, since their high radiation resistance and high dissolving ability related to elemental sulfur. As components of complex reaction systems anionic and nonionic emulsifiers were used, which physicochemical properties are given in table 1.

Table 1. Physicochemical properties of emulsifiers.

№	Name of the emusilfier	M, g/mol	ρ, g/cm^3	T_{melt}, K
1	Sodium carboxymethyl cellulose SCMC	25-30·10^3	1.59	–
2	Sodium dodecyl sulfate SDS	288.38	1.01	478
3	Sodium dioctyl sulfosuccinate SDSS	444.55	–	428
4	Triton X-100	625	1.07	279
5	Triton X-305	1526	1.09	–

Purification of sulfur was carried out by recrystallization by a standard method. The emulsion preparation procedure was as follows: 200 ml of benzene and a weighed portion of purified sulfur were placed in a 500 ml round bottom flask; the flask was heated for 1 h under reflux; the resulting solution was cooled, filtered and placed in a dark place before use.

We studied emulsions of sulfur in benzene with various emulsifiers under different conditions (table 2). When choosing an emulsifier for emulsions and subsequent polymerization under the action of high-energy radiation, into account must be taken long-term aggregative stability (in the field of radiation) and the possibility to extract the resulting product from the emulsion.

Table 2. Characteristics of studied emulsions.

№	Emulsifier	Ionizing radiation	Microwave radiation
1	SCMC	+	–
2	SDS	+	+
3	SDSS	+	+
4	Triton X-100	–	+
5	Triton X-305	–	+

To prepare emulsions, the water solutions of emulsifiers were prepared. Then, 10 ml of sulfur solution in benzene was added in a beaker with 90 ml of water solution of emulsifier, and subjected to ultrasonic treatment (UZDN-1) for 15 min. Several preliminary experiments have shown that such a time is sufficient for complete dispersion.

The emulsions were placed in test tubes and closed with lids. They were irradiated by ionizing radiation (P = 0.108 Gy/s, ^{60}Co) or by microwave radiation (2.45 GHz, Biotage Initiator+).

After irradiation, the tubes were opened, the solution was filtered, the precipitate was washed with distilled water and dried at room temperature. The synthesized products were investigated by photon correlation spectroscopy (Photocor Compact-Z spectrometer), scanning electron microscopy with elemental analysis (JSM 6510 LV SSD X-MAX), and X-ray diffractometry (Dron 3M diffractometer) on the basis of the Center for Collective Use of the Dmitry Mendeleev University of Chemical Technology of Russia.

3. Results

After the effect of the ionizing radiation on the emulsions as the products were obtained the sulfur-containing precipitates of light yellow color. After filtering the irradiated emulsions the samples of precipitates were analyzed to determine the average particle size and particle structure of precipitates. On the microphotographs (figures 1, 2) one can see that two fractions prevail in the systems. Larger particles are agglomerates, their shape is close to rhombic and monoclinic, their size varies in the range 110–170 μm. The particle size of the fine fraction is 10–50 μm. The sample from system 2 with SDS also contains particles of an amorphous-fibrous structure, which may be a sign of an oligomeric or polymeric structure. It should be noted that the particles in sample 3 with SDSS also have a shape close to rhombic and monoclinic, but, unlike the particles in sample 2, they are 10 times smaller and close to nanoscale particle, and also in sample 3 particles of an amorphous-fibrous structure are absent.

Figure 1. SEM micrograph of sulfur-containing precipitate from sample 2.
[SDS] = 1% wt., D = 2.33 kGy.

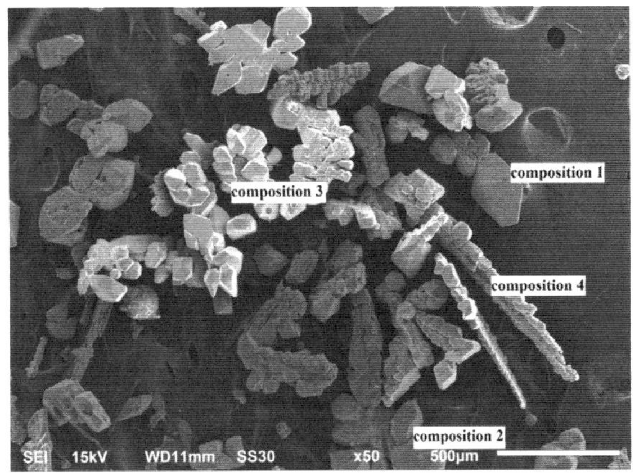

Figure 2. SEM micrograph of sulfur-containing precipitate from sample 2 (points of composition determination are indicated).
[SDSS] = 1% wt., D = 2.33 kGy.

Micrographs of particles of sulfur-containing samples confirm the theoretical idea that the polymerization proceeds in a narrow zone inside a polymer-monomer particle located on the interface, and, in accordance with this, is determined by the particle surface area, concentration and nature of the emulsifier. The depth of the reaction zone depends on the diffusion properties of macromolecules, radicals, and other components of the reaction. It is important to consider the ratio of the diffusion rates of these particles to the chain growth rates. The radicals initiating the reaction form either in the surface zone or enter the reaction zone from the aqueous phase.

Elemental analysis of the surface (table 3) showed an increased carbon content in the surface layer (34-65% wt.), as well as the presence of sodium (1% wt.). The presence of carbon in the sulfur-containing samples indicates the possibility that an organic emulsifier is involved in the formation of initiating particles.

4. Discussion

To determine the structure of sulfur-containing precipitates, X-ray diffraction was used, the results are presented in figures 3, 4. The presence of distinct peaks in the X-ray diffraction patterns of the samples indicates the crystal structure of sulfur-containing samples. However, in the regions of 60–70° there are no distinct peaks, which may indicate an amorphous structure. The observed X-ray amorphous halo is a sign of the polymer structure of the product.

Table 3. Elemental analysis of the surface of sulfur-containing samples.

Sample	Element, % wt.		
	C	S	Na
1	65	34	1
2	34	66	0
3	0	100	0

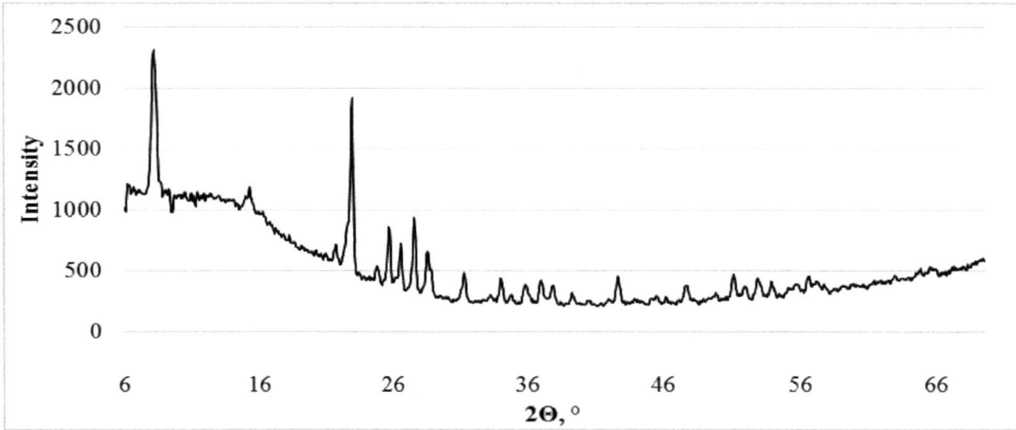

Figure 3. XRD of sulfur-containing precipitate from sample 2. [SDS] = 1% wt., D = 2.33 kGy.

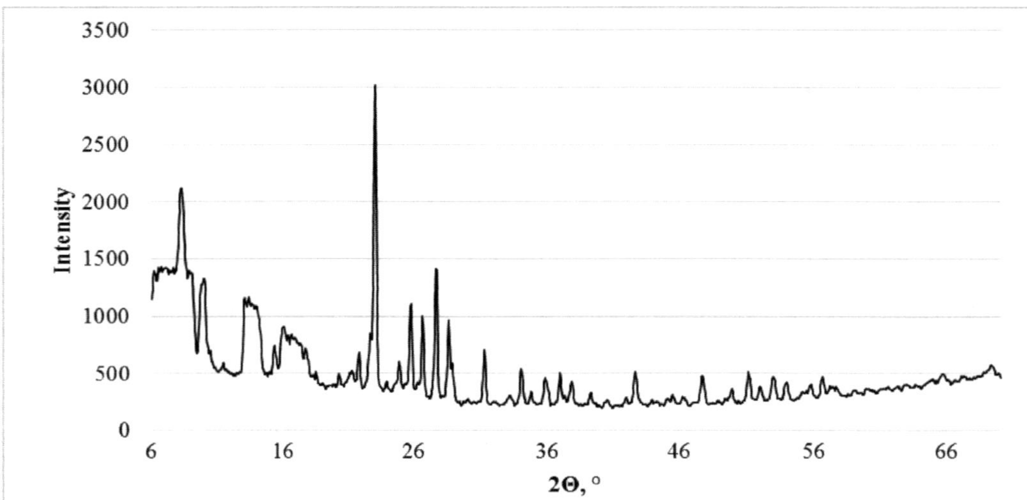

Figure 4. XRD of sulfur-containing precipitate from sample 3. [SDS] = 1% wt., D = 2.33 kGy.

The systems 2–5 were also treated by microwave radiation with the frequency 5.45 GHz. The parameters of the processes of microwave heating of the systems with ionic (SDS, system 2) and non-ionic (Triton X-100, system 4) emulsifiers are presented on the figures 5, 6, respectively.

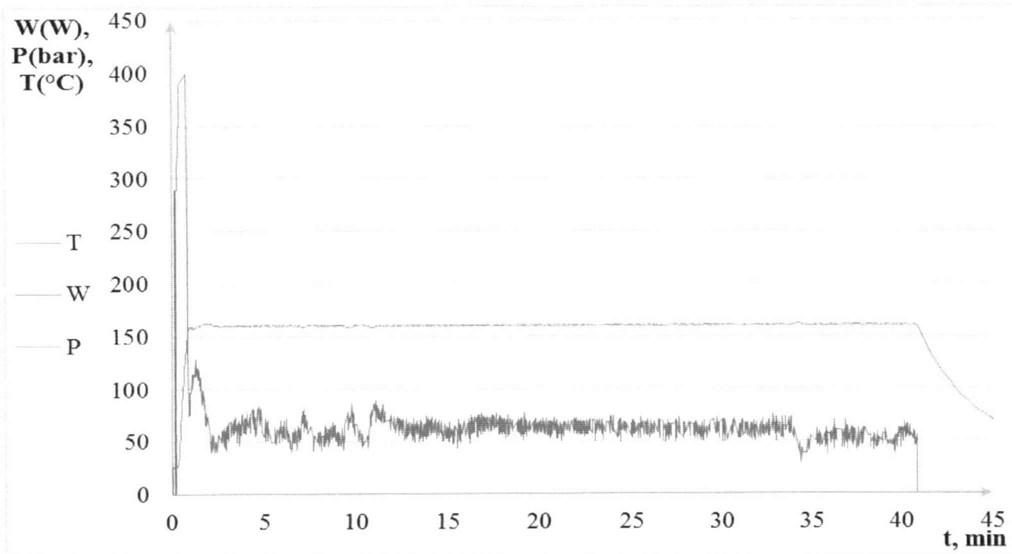

Figure 5. The microwave heating process of system 2. T = 433 K, t = 30 min, P = 2.33 bar.

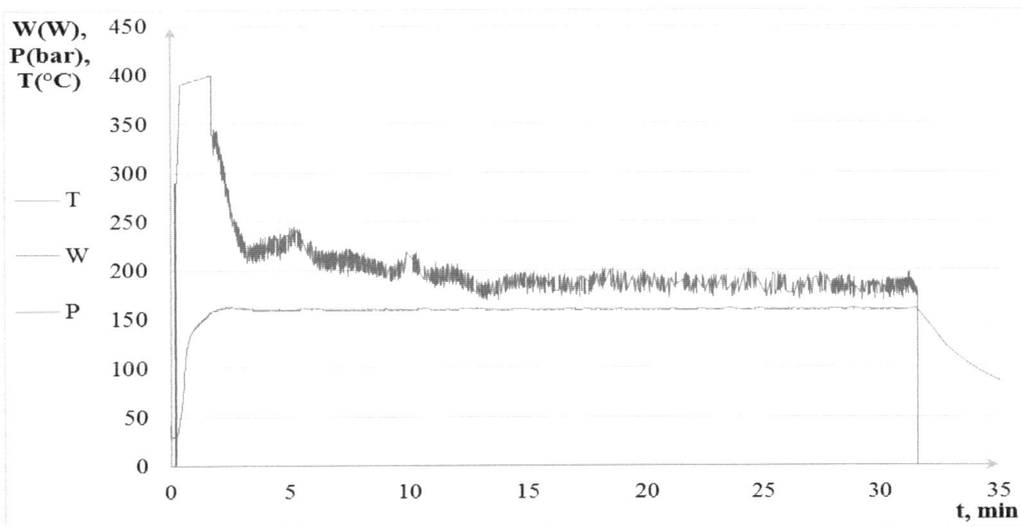

Figure 6. The microwave heating process of system 4. T = 433 K, t = 30 min, P = 8 bar.

It can be noted that the power consumption for carrying out the process with nonionic emulsifiers is about 180–200 W, while for ionic emulsifiers this value is about 50 W. This can be explained by the fact that ionic conductivity increases the efficiency of microwave heating, and lower power consumption is required to maintain the required temperature.

It is also worth noting that in the case of an ionic emulsifier, heating occurs more sharply, while the heating curve with a nonionic emulsifier is smoother. This can also be explained by the ionic conductivity factor. In systems 2-5 the precipitates of sulfur-containing granules were obtained, which shape is close to spherical.

5. Conclusion

By SEM and XRD, the structure of sulfur-containing precipitates obtained under radiation irradiation was determined. Two fractions of particles of both crystalline and amorphous structures were observed. The presence of an amorphous structure suggests that polymer (oligomer) was formed. The

XRF analysis indicates a high proportion of sulfur (66–100% wt.) and carbon (34–65% wt.) in the precipitates. The presence of carbon indicates the possibility of the participation of an organic emulsifier in the formation of initiating particles.

Using microwave radiation it was found that due to the low thermal conductivity of benzene, which is part of the emulsion, local overheating of the mixture occurs. In this regard, it was concluded that it is necessary to increase the pressure during the reaction. When carrying out the reaction at a temperature of 433.15 K and a pressure of 0.8 MPa, it was found that maintaining the temperature during the process requires less power consumption for ionogenic surfactants than for nonionic ones. This factor may be due to the fact that ionic conductivity increases the efficiency of microwave heating. This is also evidenced by a sharper heating of the emulsion based on ionic surfactant compared to nonionic. During the reaction, a sulfur-containing precipitate is formed as granules with the spherical and close to spherical shapes.

References

[1] Goethals E J 1968 Sulfur-Containing Polymers *Journal of Macromolecular Science - Reviews in Macromolecular Chemistry* **2:1** 73-144
[2] Obodovskiy I 2019 *Chemical Processes Under the Influence of Radiation* (Radiation: Elsevier) 247-55
[3] Tarasova N P and Smetannikov Y V 2011 Radiation chemical synthesis of modified phosphorus-containing polymers *Doklady Chemistry* **437** 1 53-6
[4] Min K and Matyjaszewski K 2009 Atom transfer radical polymerization in aqueous dispersed media *Central European Journal of Chemistry* **7** 657
[5] Braunecker W A and Matyjaszewski K 2007 Controlled/living radical polymerization *Progress in Polymer Science* **32** 1 93-146
[6] Lopatkin D S, Shushunova T N, Shaldina G E, Gibadullin A A and Smirnova I L 2019 Renewable and small energy development management *Journal of Physics: Conference Series* **1399** 033061
[7] Morkovkin D E, Kerimova Ch V, Dontsova O I and Gibadullin A A 2019 The formation of factors affecting the sustainable development of the generating complex of the electric power industry *Journal of Physics: Conference Series* **1399** 033042
[8] Tarasova N, Zanin A, Burdakov K and Sobolev P 2015 Ionic liquids and microwave irradiation in polymer synthesis *Polymers for Advanced Technologies* **26(7)** 687-95
[9] Zimnukhova D I, Zubkova G A, Morkovkin D E, Stroev P V and Gibadullin A A 2019 Management and development of digital technologies in the electric power industry of Russia *Journal of Physics: Conference Series* **1399** 033097
[10] Mallakpour S and Dinari M 2008 Microwave step-growth polymerization of 5-(4-methyl-2-phthalimidylpentanoylamino) isophthalic acid with different diisocyanates *Polymers for Advanced Technologies* **19** 1334-42
[11] Liao L, Zhang C and Gong S 2007 Microwave-Assisted Synthesis and Characterization of Poly(ε-caprolactone)/Montmorillonite Nanocomposites *Macromolecular Chemistry and Physics* **208** 1301
[12] Gibadullin A A, Pulyaeva V N, Kharitonova E N, Erygin Yu V and Kharitonova N A 2019 Formation of a mechanism to ensure the stability of the electric power complex *IOP Conf. Series: Materials Science and Engineering* **537** 042031
[13] Morozov G A, Morozov O G, Nasibullin A R and Samigullin R R 2011 Fizika volnovykh protsessov i radiotekhnicheskie sistemy *Macromolecular Chemistry and Physic* **14** 114
[14] Hoogenboom R and Schubert U S 2007 Microwave-Assisted Polymer Synthesis: Recent Developments in a Rapidly Expanding Field of Research *Macromolecular Rapid Communications* **28** 368-86
[15] Linnik V Yu and Linnik Yu N 2019 Priority directions of innovative development in oil and gas complex at the Siberian Federal District *Upravlenie* **7** 1 40-9

IOP Conf. Series: Materials Science and Engineering **837** (2020) 012004 doi:10.1088/1757-899X/837/1/012004

Formation of the load graph for consumers of electric energy for the stabilizing work of electric energy system

D N Udintsev, G V Shvedov, P K Milovanov and M M Sergeeva

National Research University "Moscow Power Engineering Institute", 17, Krasnokazarmennaya, Moscow, 111250, Russia

E-mail: udincevdn@mail.ru

Abstract. Design and operation of electric energy system is being carried out with taking into account probable projected accidents. Thus, there are many prerequisites of developing accidents in electric energy systems, caused by impact of outer factors, which have random nature and can negatively influence the stability of operation of electric energy system. Problematic questions of functioning of electric energy systems in accidents, caused by various factors, which have both predicted and random nature are being considered in this article. There are actual ways of preventing development of accidents provided. A method of flexible regulation of consumption of electric energy had been suggested as one of the possible ways to stabilize the operation of electric energy system.

1. Introduction

Nowadays, the consumption of electric energy all over the world has a positive dynamic. The growth of consumption is provided by demand for electric energy in developing economics. So the small stagnation of electricity consumption of USA, Europe and Russia in crisis years has been fully compensated by the growth of electricity consumption in China, India and developing countries of Asia-Pacific region. In whole, in Russia the dynamic of consumption fits world trends of growth and decrease but the volume of production is stable.

The actual consumption of electric energy [1-4] in Russian Federation is growing in recent years (table 1).

Table 1. Consumption of electric energy in UES of Russia.

Year	Consumption, billions of kW per hour	Gain regarding the previous year, %
2015	1008.2	-
2016	1026.7	1.8
2017	1039.7	1.3
2018	1055.6	1.5

Firstly, growth of the consumption of electric energy is caused by increasing of production on the functioning industrial enterprises and also by commissioning the new factories. Mostly this gain is produced by factories of woodworking industry, metallurgical factories, and also by both oil and gas pipeline facilities and railway transport. Moreover, a household sector contributes to a growth of electricity consumption – therefore the power ratio of personal households is growing, very new consumers and receivers of electric energy are appearing, rapid development of personal electric transport starts.

The installed capacity of Russian power plants increases every year [4], generally responding to the growing electric consumption. United energy system of Russia shows quite high stability of operation. Every year the number of accidents decreases [1]. At the same time, with the general decrease in the number of accidents at electricity facilities in general and the decrease in the number of major system accidents in particular, their scale becomes larger and they develop according to the cascade scenario.

2. Materials and methods

In the emergency and post-emergency conditions, the system is at a lower level of operation than normal operation. Any minor disturbance causing a decrease in the level of functioning with a high probability can turn into a cascade development of the accident and cause a mass disruption of the power supply to consumers. As systems become more complex, the number and severity of cascading system accidents increases, as well as the part of power shortage caused by these accidents increases. The reason is not only in the increase of the total number of dependent failures, but also in the complication of control in accidents, the difficulty of setting up emergency automation. The necessity and importance in ensuring the vitality of the electric power system, the reliability and quality of electric power supply to electric power users, as well as the power shortage reduces, makes it particularly relevant in the development of ways to prevent the development of accidents and the stable functioning of the electric power system with the impact of outer unfavorable factors of accidents.

In this regard, the aim of the research is to adapt the operation of the electric power system to accidents, resulting from negative outer factors.

To achieve this goal, the following tasks must be solved:

- To investigate features of functioning of an electrical power system in the accidents resulting from action of negative outer factors;
- To analyze the existing and perspective ways of prevention of accidents;
- To offer a possible way of stabilization of operation of an electrical power system.

The study was based on methods of statistical, historical, comparative, logical, economic-mathematical and system analysis, a method of expert assessments, which allowed to propose a method of flexible regulation of electricity consumption, as one of the possible ways to stabilize the operation of the electric power system.

3. Results

The analysis of daily reports of the Federal State Budget Institution "Situation and Analytical Center of the Ministry of Energy of Russia" on accidents that occurred at the facilities of the fuel and energy complex [5] shows that the occurrence of accidents of different scale and duration is coming out due to a number of main reasons.

The first reason is the occurrence of accidents at electric power facilities as a result of unfavorable weather conditions (heavy rain, wind gusts, storm). There is a large number of prerequisites for the occurrence of negative natural impacts, which are random and can negatively impact the operation of the elements of the electric power system. The territory of the Russian Federation is exposed to almost the entire spectrum of natural hazards and processes. The climate is changing rapidly, leading to widespread natural disasters. Among possible natural phenomena, earthquakes present the greatest risk (about 20% of the territory of Russia are exposed to earthquakes with intensity of more than 7 points) and floods (they occupy the first place in Russia among dangerous hydrological phenomena and processes in terms of area of spread and total average annual damage, and the second, after earthquakes, in terms of the number of victims and damage). The second reason comes due to the operation beyond the standard service life of more than half of the electric power equipment in operation. According to the data of the technical inspection of EEC [6], the average service life of the main generating equipment of power plants in Russia in whole is:

- The steam turbines, gas turbines, water-wheels installed on power generation facilities which total established generating power in total exceeds 5 MW - 32.1 years;
- The package boilers providing the ferry the steam turbines installed on power generation facilities which total established generating power in total exceeds 5 MW - 44.7 years;
- The turbo generators and hydro generators installed on power generation facilities - 32.8 years;
- Transformers (autotransformers) power oil, established on power generation facilities, a class of voltage of 35 kV and above - 28.3 years.
- The average service life of the main electrical equipment, air and cable lines (hereinafter - power lines) of electrical networks in Russia in whole is:
- The transformers installed on objects of electrical networks, a class of voltage of 35 kV and above - 31.7 years;
- The power line a class of voltage of 35 kV and above - 39.7 years.

In 2019, 2969.9 MW of new capacity was introduced at power plants (including industrial power plants). 1746.03 inefficient and obsolete equipment has been decommissioned. At the same moment, the operation of a large number of equipment with unacceptably high wear and tear continues, operating at the border or outside the standard service life [2].

The next frequent reason of accidents is incorrect or false operation of emergency automation (EA). The fault is the technical incompatibility of devices of different manufacturers and inconsistency of algorithms of their operation. Besides, there are errors in adjustment of relay protection (RP) - protection settings do not take into account the possibility of operation of network elements with additional load, and its settings do not always meet the requirements under maximum loads and emergency modes. Failure of one element is accompanied by disconnection of the protection means of the network element taking over its load, which leads to cascade development of the accident and mass disruption of the consumers supply. In general, the reasons of accidents in electrical power systems can be conditionally divided into organizational, technical and accidents which have a random nature (figure 1).

There is a seasonal change in the number of accidents at electric power facilities of different voltage classes in the section of the year [3]. Seasonal accidents can be predicted, what helps to minimize their consequences by carrying out preventive measures by electricity grid organizations and utilities and bringing the forces and facilities of repair teams to full readiness. Accidents which have a random nature are not predictable and become an exceptional situation. The reasons of such accidents may come the effects of natural emergencies that are not typical for the region or have not previously occurred with such intensity. In addition, the current level of industrial development should involve a serious approach to assessing the man-made hazards that may arise in the operation of complex industries and technologies. One of the reasons for the accidents is the violation of the security zones of power transmission lines in the part of the non-coordinated construction works near the electric power infrastructure, non-compliance with the vertical dimensions of construction and heavy equipment when passing under the overhead line (OL) of power transmission. There are cases of light-powered aircraft, quadrocopters, drones and balloons falling on the OL. Vandalism, as well as deliberate deactivation of electrical network elements, is a rare but at the same time severe cause of accidents at electric power facilities.

The problem of operating a large number of electric power equipment outside the standard service life is exacerbated by the poor technical condition of the reserve equipment. This circumstance does not allow to put the reserve equipment into operation quickly in case of accidents. In addition, when new equipment is put into operation, there is often a problem of non-conformity of technical characteristics of the installed equipment with the parameters of that, which was already in operation.

IOP Conf. Series: Materials Science and Engineering **837** (2020) 012004 doi:10.1088/1757-899X/837/1/012004

All these lead to their inconsistent operation, incorrect adjustment of protection algorithms and failures.

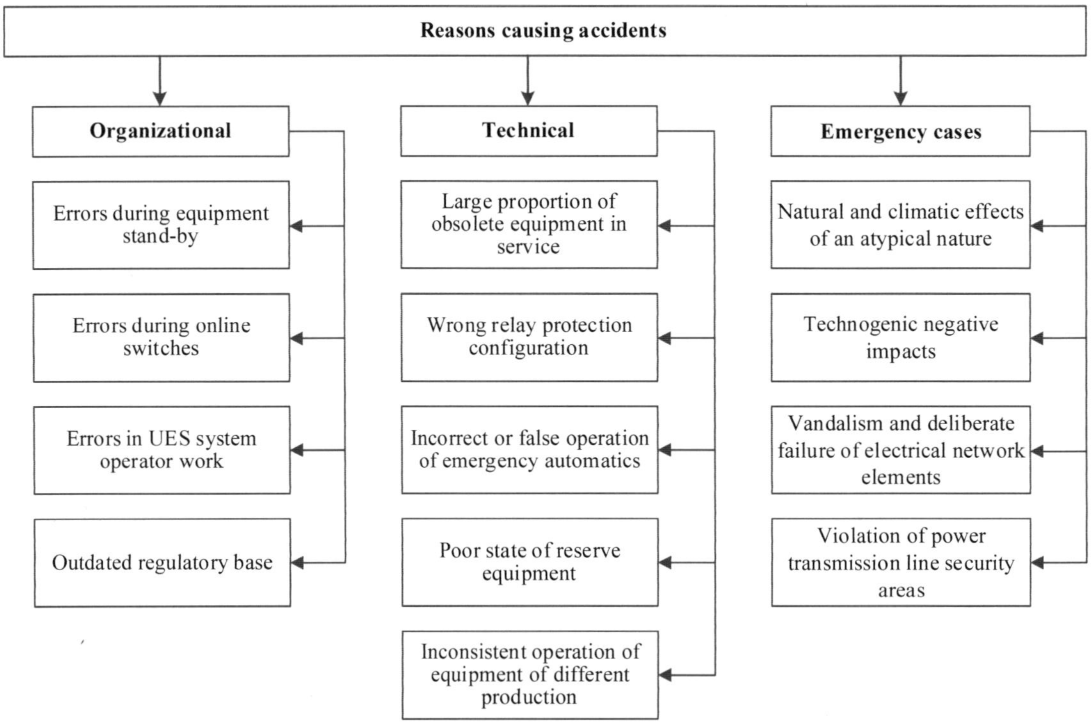

Figure 1. Reasons, causing acciden.

A lot of the reasons of system accidents are coming due to organizational work. The situation is complicated by insufficient qualification of personnel and their errors in operational switching or interference with relay protection and automation. The organization of network and generating equipment outputs for repair is not always sufficiently thought out. This results in reduced reliability of power systems, especially in seasonal peak load conditions. There are also cases of inconsistent interaction of the system operator with electric power industry entities in terms of protection configuration, issuance of tasks and control of their execution.

The main direction of development of emergency control in the UES of Russia is improvement of emergency automation systems [7].

The emergency control solves the problems: prevention of power system operation stability failure, termination of asynchronous stroke, prevention of frequency, voltage and current exceeding permissible limits.

Modern types of emergency automation devices used in the operation of the UES of Russia are presented in table 2.

According to the regulatory documents [8-9], emergency control functions are implemented by AP by means of the following control actions:

- Short-term (pulse) and long unloading of power units of thermal power plant and NPP;
- Shutdown of generators;
- Shutdown of load of consumers of electric energy;
- Forsirovka of excitement of generators;
- Division of power supply system into nonsynchronously working parts;
- Automatic loading of generators;

- Electric braking;
- Change of topology of an electrical network;
- Change of operating modes and an operational condition of the operated elements of an electrical network.

Table 2. Kinds of emergency automation devices.

A task, solved by an emergency automation	Kind of EA	Purpose
The prevention of stability violation	SFPA Stability Failure Prevention Automation	Prevention of violation of static and dynamic stability of generating equipment of power plant, engine load assembly, controlled section, power area and prevention of unacceptable current overloads of power lines and equipment.
Terminations of the asynchronous course	AATE Automatic Asynchronous Travel Elimination	Detection and elimination of asynchronous modes of individual generators, power plants and parts of power systems, detection and elimination of asynchronous full-phase and incomplete-phase modes of electric network.
The prevention of invalid frequency deviations	AFIL Automatic frequency increase limitation	Prevention of unacceptable increase of frequency in the power system to the level, at which it is possible to activate the safety automatics of power plant and NPP turbines.
	AFRL Automatic frequency reduction limitation	Prevention of frequency reduction and complete redemption of the power system or its part in case of active power deficit, including emergency allocation of the power system or its part for isolated operation, which is unacceptable by conditions of stable operation of generating equipment and power receiving plants of electric power consumers
Prevention of unacceptable voltage deviation	AVIL Automatic voltage increase limitation	Prevention of unacceptable value and duration of voltage increase on equipment of electric power industry objects.
	AVRL Automatic voltage reduction limitation	Prevention of voltage reduction unacceptable in terms of stability conditions of generating equipment and power receiving plants of electric power consumers.
Prevention of an unacceptable overloading of equipment	EOLA Equipment overload limitation automation	Prevention of unacceptable current load value and duration of electrical equipment and power line.

One of the most common control actions of emergency automatics designed to prevent the development of emergency modes in the EES is to disable the load [10]. Disconnection of load (DL) of electric power consumers is used to prevent stability violations, limit reduction of frequency and voltage, eliminate overload of controlled sections, power lines and equipment. DL is performed by disconnection of all electrical connections of power receiving plants of electric power consumers with power system with prohibition of automatic entry of reserve of disconnected connections. At the same time, this effect is the least desirable because it reduces the main indicator of reliability of power supply to consumers. Both concentrated energy-intensive consumers and large distribution centers can be connected under DL. At the same time, in today 's conditions, it is almost impossible to abandon the DL in many cases, especially to solve the problems of sustainability of inter-system ties of the UES.

4. Discussion

The issue of stabilization of the operation of the electric power system, as well as preservation of the stability of its operation in case of power imbalances arising from accidents, is devoted to a lot of works of domestic and foreign authors.

One of the solutions to this issue can become the method of determining the value and placement in the system of reserve for consumers (power of short-term shut-down load) necessary to prevent the development of an accident caused by the shutdown of a large electric station in the EES [11].

An another way to prevent emergency situations can be to place the consumer-regulator in the system [12]. It is a consumer, whose technological process and mode of operation on a reimbursable contractual basis provide for the possibility to limit power consumption during the hours of maximum load of the electric power plant and increase the load during the hours of minimum. These measures are necessary to balance the daily load schedule of the power system.

There are developments on construction of adaptive automation of distributed load disconnection, as well as algorithms of formation of adaptive control actions in order to ensure stability of EES [13].

Currently, developments are under way to introduce Intelligent Energy Systems (IES) or Smart Grid. One such system is an intelligent system based on multi-agent technologies and machine learning algorithms for early warning and emergency prevention [14-18].

In 2011, at a joint meeting of the NTS of JSC "FSK UES" and the Russian Academy of Sciences, the Concept of Development of the Intelligent Electric Power System of Russia with an Active and Adaptive Network (IES AAN) was approved [16]. The intelligent electric power system with active-adaptive network is a new generation electric power system based on the multi-agent principle of organization and management of its functioning and development, based on modern technological means and a single intelligent hierarchical control system.

Load balancing is one of the factors of stable operation of the electric power system. One way to optimize daily load schedules is to use market mechanisms and develop digital technologies while encouraging consumers to actively manage their own consumption [19-21]. In this case, it is necessary to take into account the mismatch of interests of the power supply organization and consumers in the process of load balancing. In addition, the development of load schedules with the use of batteries as consumer regulators is under way [22].

Most existing approaches to maintaining the stability of the electric power system in case of power imbalances by switching off the load do not take into account the presence of different requirements for the reliability of electric power supply for individual consumers. In this regard, it is necessary to develop a method of stabilizing the operation of the electric power system, taking into account the requirements for reliability of electric power consumers and excluding their complete shutdown.

In order to solve this issue, the authors of the article propose flexible regulation of electricity consumption. [23] The main essence of this method is the possibility to generate a load graph, depending on the availability of free power in the system, taking into account the capacity of the electrical network elements.

The ability to generate a graph is achieved by:

- Temporary increase in capacity of a system of transfer and power distribution due to accounting of total thermal loading of separate elements of networks (transformers, air and cable lines);
- Transition from discrete regulation (shutdown of consumers of the electric power) to smooth (change of value of settings of devices of protection and the relay of priority loadings) on low-priority consumers of the electric power during the periods of lack of free power in a system.

- Preliminary start of reserve power sources of consumers of the I category of reliability of special group and transfer on them, as on the main source of the electric power, the specified consumers.

An example of load graph generation is shown in figure 2.

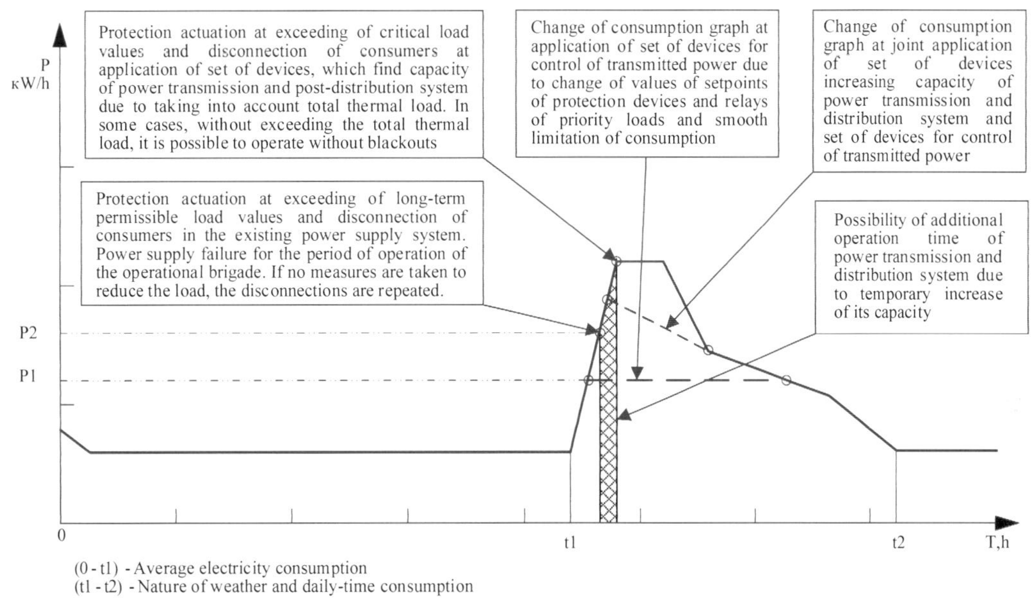

Figure 2. The example of load graph generation.

Thus, load parameters are adapted to the generation capabilities and capacity of the electrical network.

At present, in order to implement the proposed method, the authors of the article are conducting studies aimed at preventing the development of emergency situations in the electric power system by using a set of technical means that provide flexible regulation of electricity consumption.

The set of technical means developed in the framework of this study [23-25] consists of several sets of devices, which are based on unique technical proposals to prevent the development of accidents:

- Set of devices increasing capacity of power transmission and distribution system due to taking into account total thermal load of individual network elements;
- Set of transmission power control devices in distribution networks in case of power shortage in the power system.

The proposed method of flexible regulation of electric power consumption can be used not only to prevent the development of emergency situations in the existing power system, but also for isolated power systems and power systems containing distributed generation to adjust the load schedule to the available free power in the system.

5. Conclusion

The article considers problematic issues of operation of electric power systems in emergency situations caused by various factors, which are both predictable and accidental. The current ways to prevent the development of emergency situations are presented. The authors propose a method of

IOP Conf. Series: Materials Science and Engineering **837** (2020) 012004 doi:10.1088/1757-899X/837/1/012004

flexible regulation of electric power consumption as one of the possible ways to stabilize the operation of the electric power system.

References

[1] *Ministry of Energy of Russia: Emergency* Retrieved from: https://minenergo.gov.ru/node/267

[2] *Russian Technical Supervision Energy: Equipment* Retrieved from: http://www.gosnadzor.ru/activity/analiz/energy

[3] *Federal State Budget Institution "Situation and Analytical Center of the Ministry of Energy of Russia": Emergency situations* Retrieved from: https://xn--80agflthakqd0d1e.xn--p1ai/map/Pages/emergency_situations.aspx

[4] *System operator of the UES: Activities* Retrieved from: https://so-ups.ru/index.php?id=ees

[5] *Ministry of Emergency Situations of Russia: Operational information* Retrieved from: http://www.mchs.gov.ru/operationalpage

[6] *Technical inspection of UES: Assessment of technical condition of power plants and networks equipment (by operating time and service life)* Retrieved from: http://www.ti-ees.ru/functioning/analytics/time

[7] *System operator of the UES: Goals and objectives of emergency management* Retrieved from: http://so-ups.ru/index.php?id=rza_goals

[8] *System operator of the UES: Standards, rules, norms and requirements* Retrieved from: http://so-ups.ru/?id=1090

[9] *ROSSETI: Legal framework* Retrieved from: http://www.rosseti.ru/clients/baza

[10] Ovcharenko N I 2009 *Automation of power systems* (Moscow: MPEI Publishing House) 311

[11] Dubitsky M A 2011 Safety of power systems at emergency switching-off of large power plants *Modern technologies. System analysis. Modeling* **2(30)** 174-8

[12] Voropay N I, Kovalev G F, Kucherov Yu N et al. 2013 *The concept of ensuring reliability in the electric power industry* (Moscow: Publishing House Energia) 104

[13] Voropay N I, Yefimov D N, Karatayev B N et al. 2016 Adaptive algorithms for automatic distributed load shedding *Electrical stations* **11(1024)** 27-35

[14] Voropay N I, Kurbatskiy V G, Tomin N V et al. 2016 *Complex of intelligent means of preventing major accidents in electric power systems* (Novosibirsk: Publishing House Science) 183

[15] Konovalov Yu V and Kuznetsova N V 2017 Improving operation safety of electrical complexes of power supply systems under their intellectualization *Proceedings of Irkutsk State Technical University* **1(120)** 103-12

[16] Berdnikov R N, Dementyev Yu A, Morzhin Yu I and Shakaryan Yu G 2012 Concept of Development of Intelligent Electric Power System of Russia with Active Adaptive Network *Unified Network Energy* **4(4)** 4-11

[17] Hidayatullah N A, Kalam A and Stojcevski B 2011 Analysis of Distributed Generation Systems, Smart Grid Technologies and Future Motivators Influencing Change in the Electricity Sector *Smart Grid and Renewable Energy* **2(2)** 216-29

[18] Afgan N H and Cvetinovic D B 2011 Resilience of High Voltage Transmission System *Energy and Power Engineering* **03(05)** 600-6

[19] Morkovkin D, Lopatkin D, Sadriddinov M, Shushunova T, Gibadullin A and Golikova O 2020 Assessment of innovation activity in the countries of the world *E3S Web of Conferences* **157** 04015

[20] Pulyaeva V N, Zlotnikova G K, Gibadullin A A, Romanova Ju A and Yuryeva A A 2019 The development of the logistics system of the electric power complex *IOP Conference Series: Materials Science and Engineering* **537** 042033

[21] Eisenberg N I, Stashevich E V and Voropai N I 2016 Coordination of cooperation of power supply company and active consumersin the optimization of daily load curves *Proceedings of the RAS. Energetics* **3** 15-25

IOP Conf. Series: Materials Science and Engineering **837** (2020) 012004 doi:10.1088/1757-899X/837/1/012004

[22] Shklyarskiy Ya E and Bragin A A 2012 Rational formation of electric load schedule of the electrical complex of the mining enterprise *Journal of Mining Institute* **196** 281-4
[23] Udintsev D N, Shvedov G V and Milovanov P K 2019 Prevention of development of emergency situations in the electric power system through flexible regulation of electricity consumption *Electricity. Transmission and distribution* **5(56)** 58-63
[24] Morkovkin D E, Kerimova Ch V, Dontsova O I and Gibadullin A A 2019 The formation of factors affecting the sustainable development of the generating complex of the electric power industry *Journal of Physics: Conference Series* **1399** 033042
[25] Gibadullin A A, Sadriddinov M I, Kurbonova Z M, Shedko Yu N and Shamraeva V V 2020 Assessment of factors ensuring sustainable development of the electric power industry in the context of transition to renewable energy sources of the national economy *IOP Conference Series: Earth and Environmental Science* **421** 032051

IOP Conf. Series: Materials Science and Engineering **837** (2020) 012005 doi:10.1088/1757-899X/837/1/012005

New approach to the implementation of the under frequency load shedding function in micro grid

Borbad G. Mirzoevich[1], Makhmadi S. Makhmudovich[1], Sirojiddin R. Chorshanbiev[1] and Rahimjon H. Saidzoda[2]

[1] Tajik Technical University named after academician M.S.Osimi, 10, academic Rajabov's avenue, Dushanbe, 734042, Tajikistan
[2] Branch of the National Research Technological University "MISiS" in Dushanbe, 7, Nazarshoeva Street, Dushanbe, 734042, Tajikistan

E-mail: borbadtjk@mail.ru

Abstract. In the presented study, a modern method has been developed related to the analysis of the network at 10 kV and 0.4 kV, using the methods of mathematical modeling it is possible to ensure the creation of the same network with different parameters. It can be seen from the work that the UFLS teams make it possible to ensure the stability of the functioning of the electric power system.

1. Introduction

The development of the electricity market today and especially in the long term requires a detailed consideration of the energy dependence of modern infrastructure, information systems and socially significant facilities during the implementation of emergency load shedding. For such disconnections are used the devices of special automation of load limitation (SALL) [1; 6], under frequency load shedding (UFLS) and dispatcher commands. When the load is disconnected at a level of 0.4 kV, all consumers are switched off, however, with selective disconnection of a load at a level of 0.4 kV, can be saved some important loads. When the load is disconnected at a level of 0.4 kV, all consumers are switched off, however, with selective disconnection of a load at a level of 0.4 kV, can be saved some important loads. In power systems, an emergency limitation during frequency reduction is implemented by the multi-element structure of software and hardware.

When the micro grid is switching to isolation mode the reliability of power supply of important consumers is providing by translate the load shedding function from the 10kV to 0.4 kV. The transition to the level of 0.4 kV is achieved through the integration of the Automated System for Commercial Accounting of Electric Power (ASCAEP) and the emergency automation devices, including for the implementation of UFLS commands [7].

The main purpose of the fast-action of UFLS's settings is to prevent a decrease in frequency below 46 Hz. and compliance of the conditions of selectivity, i.e. avoiding unnecessary load shedding with subsequent increasing frequency above 50 Hz. The UFLS-I is considered in this paper.

2. Materials and methods

The aim of this work is to study the approach to implementing the load disconnect function at a low frequency. The objectives of the study are as follows:

- Study management systems;
- Simulate the management process.

The study is based on the use of various scientific methods and technical approaches, which allowed forming a model of the control system.

3. Results

An example of the integration structure of the programming technical complex (PTC) of UFLS and ASCAEP shows in Fig 1. There is envisaged to maintain the current order of location of USLS devices on 10 kV feeders with an action to trip. When the frequency decreases, the command from the UFLS device is transmitted to the 10/0.4 kV network for disconnecting the load in the form of a dip of the supply voltage for a given duration (taking into account the scatter) by disconnecting the on / off switch of the 10 kV feeder's breaker. The management of each of the disabled 0.4 kV consumers is carried out by the contacts of individual electricity metering devices. In this case in many of the electricity counters will start a control algorithm for metering the duration of the dip voltage, subsequent activation after a specified time of the circuit breaker 10 kV generates a command to trip the network of 0,4 kV, which reacting many of the energy-accounting counters. After that, the most prioritized (important) part of the electric consumers will not switch off, which it is the purpose of this work.

The output circuit of the UFLS device is connected to the logical input of the microprocessor protection relay, which will turn off the breaker without delay (figure 1). This implementing a variant of the fast-acting load shedding by command of settings UFLS-I.

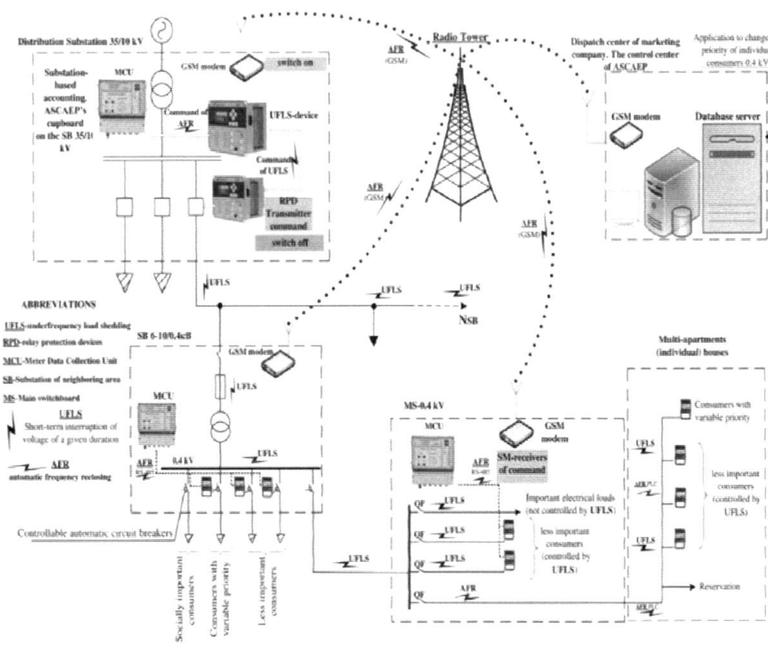

Figure 1. The structure of integration of software-hardware complex of UFLS and ASCAEP's systems.

The software and hardware complex [2-3] is proposed to be implemented by integration with existing microprocessor automation and protection relay (PRA) on 10 kV level and programmable electricity metering devices with a control contact at consumers 0.4 kV. The UFLS devices currently installed remain with their position on the 10 kV level, and their tripping signals are transmitted to the 0.4 kV network using the switching cycle (on/off) of the 10 kV breaker. Signals for the implementation of control actions are distributed in the 0.4 kV network in the form of voltage dip with a pre-determined duration and are implemented in the algorithms of ASCAEP's meters to turn off consumers. Accordingly, in the unified energy system will formed the unified management system [4-5].

Using the structure proposed in figure 1, it can be used for load shedding when occurs overload and turn it on again. The use of ASKAEP devices is considered effective. However, in case of power system trouble their operation is considered unreliable [5]. The impact of induction motors on the duration of voltage dip which uses as a signal was studied in [6; 8-11].

4. Discussion

The behavior of the power system's frequency change under the action of the UFLS obtained by modeling in PSCAD software shows in figure 2 under the following conditions.

The maximum calculated deficiency of active power during the transition to the island mode is $\Delta P_{def.max}$ = 50%, then the maximum amount of the load connected to the fast action UFLS-I settings is accepted ΔP_{Σ} = 55% more 5%.

The inertial constant of rotating masses of power system T_J = 5.0 s, is the load-damping constant k_{reg} = 2.0. The number of settings is assumed to be 20 with frequency settings from 48.8 till 46.9 Hz. with a uniform step in frequency of 0.1 Hz and uniform distribution of disconnecting load. Each setting disconnects the load in the amount of 55/20 = 2.75%. The group of graphics is generated by one parameter of the delay time Δt_{UFLS} to implement the frequency relay signal in each of the UFLS devices. In addition, an additional delay is usually provided for detuning from short-term trips of the frequency relay during switching in distribution networks. In a series of experiments, was used a set of 3 values Δt_{UFLS} = 0.15, 0.2, 0.3s. (figures 2, 3).

The mathematical model based on PSCAD set values of the above parameters micro grid was implemented the multi-interval exponential processes. In the initial operation at t0 = 0 the time constant of the frequency change is:

$$\tau_{f0} = \frac{TJ}{K_{reg}} \tag{1}$$

where k_{reg} is the load-damping constant. When a deficiency ΔP_{def} (%) occurs, the frequency as exponential with a time constant decreases to a new steady-state frequency value:

$$f_{st.} = 50 - \frac{\Delta P_{def} \cdot f_0}{100 \cdot k_{reg}}, \tag{2}$$

Under the action of the UFLS-I in the process of reducing of frequency, a multiple reassignment of all parameters of the transient process occurs, which leads to a steady-state frequency value taking into account the action of the N-settings of the UFLS-I in the expression:

$$f_{st} = 50 - \frac{(\Delta P_{def} - \sum_{i=1}^{N} \Delta P_{1-i}) \cdot f_0}{(100 - \sum_{i=1}^{N} \Delta P_{1-i}) \cdot k_{reg}} \tag{3}$$

where ΔP_{def} - volume of power deficiency; N-is the number of UFLS-I's setting, f_0 is the nominal frequency value, and ΔP_{1-i} - the power of the i-th UFLS-I's setting.

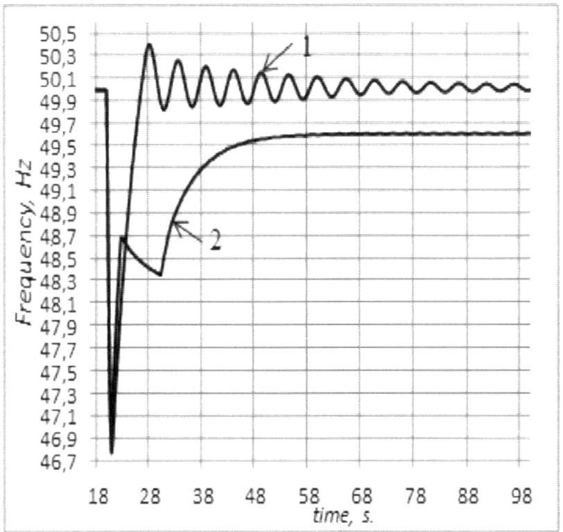

Figure 2. Frequency changing behavior when the deficiency occurs in one step. Traditional -1 and new approach-2.

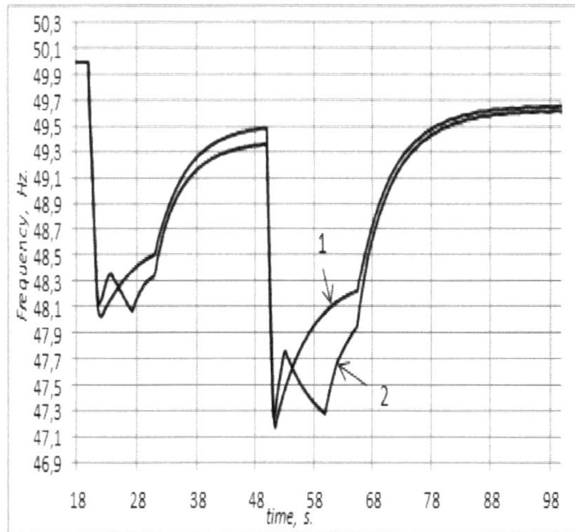

Figure 3. Frequency changing behavior when the deficiency occurs in two steps. Traditional -1 and new approach-2.

In figures 2 and 3 shows the oscillograms of Implemented model when operated the settings from 15 till 20 which formed as a set of 30-40 or more time intervals. The event (deficiency of power) will be occurs at 20 second of modeling process.

A series of graphics in figure 2 allows to visually evaluating the efficiency of the UFLS with various probable initial parameters. It is clearly seen that at Δt_{UFLS} values of 0.3 s and occurs the effect of deregulation, that is, excessive load disconnections with a subsequent increasing the frequency above 50 Hz and the action of regulators of power plant units for reducing the power generation. This situation is not permissible and is characterized as non-selective and this non-selective action is occurs due to a delay during implementation of the UFLS's signals. Authors using the duration of voltage dip as a signal in the 0.4 kV network as a task to evaluate the impact of this voltage dip at functioning of the load shedding and prevent non-selective action UFLS-I. In figure 3 shows model oscillograms of a frequency change under the same conditions as in figure 2, but in cases event of a deficit in two steps.

In this context of the implementation of the transferring of the UFLS's signals to the level of 0.4 kV used of a voltage dip with 2.0 s duration as a signal to load shedding.

The value of non-disconnecting loads on each of feeder was accepted 20%. Consequently the amount of load which to be connected to each of UFLS's setting must be increased from the value of 2.75 till 3.44%.

During operation main of breaker the feeder of 10 kV without delay will disabled with a volume of load 3.44%, and after 2.0 s. will turns on with a load of 0.68%. In this case, the 2.75% of load is disconnected by the signals of numerous meters in the 0.4 kV network. The compared oscillograms it is clearly seen that in both cases the main purpose of the fast-action of UFLS-I subsystem is quite satisfactory, with almost identical minimum frequency values $f(t)_{min}$ = 46.8 Hz. engineering

In the unified energy system of the Republic of Tajikistan the minimum acceptable frequency value is 46 Hz and under the considered conditions this requirement with a reserve in both cases is performed. But the requirement of selectivity and preventing unnecessary load shedding is performed differently.

In the traditional case, the violation of the selectivity requirement at Δt_{UFLS} = 0.3 s, according to the conditions of figure 2.

Load disconnection at 0.4 kV level under action of UFLS signals in the form of dip voltage duration 2.0 s. radically will change the transition process due to short-term load shedding and after that occurring switching on of load. Therefore, raising the frequency and the possibility of excessive non-selective load shedding on the oscillograms shows in figure 3 is not observed.

As is shown in figure 3 frequency hanging in the range of 48.5 ÷ 46 Hz. is limited in time (90-100) s. and it is eliminated by the UFLS-II settings combined with the UFLS-I settings at the same consumers. It is advisable to keep the implementation of the commands (signals) of the UFLS-II category as the settings of the UFLS-I with the transfer they implementation to the level of 0.4 kV.

In figure 4 shows the behavior of the frequency change when the deficiency is 50% and increase slowly during 4 minutes. The traditional and proposed methods as a whole demonstrate the spasmodic form of the frequency, moreover, both methods successfully prevent an "avalanche of frequency", however, they can allow an excess of the residence time duration of frequencies outside of the control area. To prevent this drawback should be in reasonable cases apply the under frequency load shedding device with time-variable setting in frequency [2; 12-14].

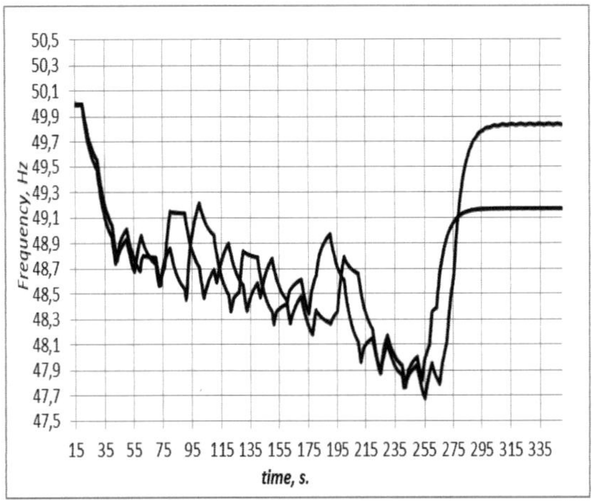

Figure 4. Frequency changing behavior when the deficiency is 50% and increase slowly during 4 minutes. Traditional -1 and new approach-2.

5. Conclusion

The simulation results clearly demonstrate the validity of the proposed method for implementing of load shedding signals to disconnection the power consumers of 0.4 kV. By the use of voltage dips as a signal to turn off power consumers the efficiency of load shedding is not reduced.

In some cases, the effectiveness of the proposed method with dip voltage can be increased by eliminating unnecessary load shedding in maximum deficits mode. This property will increase in the future as the influence of the inertia of the rotating parts of the power system decreases by using renewable energy sources which operating with invertors.

At subsequent stages of the research it is advisable to move the pilot project and to equip a feeder for the 10 kV devices for transmission signal to load shedding in the 0.4 kV network.

References

[1] Zimnukhova D I et al 2019 *J. Phys.: Conf. Ser.* **1399** 033097
[2] *Enabling Tomorrow's Electricity System: Report of the Ontario Smart Grid Forum* Retrieved from: http://www.ieso.ca/imoweb/pubs/smart_grid/Smart_Grid_Forum-Report.pdf
[3] Gibadullin A A, Pulyaeva V N and Yerygin Y V 2018 The Need for a Digital Substation during the Digitalization of Energy *International Youth Scientific and Technical Conference Relay Protection and Automation RPA* 8537223
[4] Qi W, Yi T and Li F 2016 Coordinated Scheme of Under-Frequency Load Shedding with Intelligent Appliances in a Cyber Physical Power System *Energies* **9** 630
[5] Taha A R and Tarik A A 2010 Wireless Distributed Load-Shedding Management System for Non-Emergency *Cases. World Academy of Science, Engineering and Technology International Journal of Electrical, Computer, Energetic, Electronic and Communication*

IOP Conf. Series: Materials Science and Engineering **837** (2020) 012005 doi:10.1088/1757-899X/837/1/012005

Engineering **4** 2 835
[6] Krzysztof B 2015 Smart Meters and Under-Frequency Load Shedding *Journal of Energy and Power Engineering* **9** 905-11
[7] Rabinovich R S 1989 *Automatic underfrequency load shedding of power systems* (Moscow: Energoatomizdat) 352
[8] Artsishevsky Y L and Giyoev B M 2015 Reliability of power supply of responsible consumers 0.4 kV at action of AUFLS *International Scientific and Technical Conference "Prospects for the develop-ment of electricity and high-voltage electrical equipment, switching apparatus, conversion techno-logy, microprocessor control and Protection Systems* **1** 247-51
[9] Artsishevsky Y L and Giyoev B M 2017 Efficiency of Under-Frequency Load Shedding with the 0.4 kV Current-Using Equipment Tripping Commands Transmission *Izvestiya Vysshikh Uchebnykh Zavedenii. Elektromekhanika* **60** 6 37-44
[10] Sterninson L D 1975 *Transients in the regulation of frequency and power in power systems* (Moscow: Energy) 216
[11] Giyoev B and Artsishevsky Y 2018 Load Shedding Devices in Distributed Generation: A *Review International Scientific-Technical Conference on Actual Problems of Electronic Instrument Engineering* **1** 119-23
[12] Lopatkin D S et al 2019 *J. Phys.: Conf. Ser.* **1399** 033061
[13] Giyoev B and Artsishevsky Y 2018 A method for Selecting the Parameters of Emergency Load Shedding Commands in 0.4 kV Networks *Bulletin of the Moscow power engineering institute* **3** 21-8
[14] Morkovkin D E et al 2019 *J. Phys.: Conf. Ser.* **1399** 033042
[15] Artsishevsky Y L and Chimaev R V 2005 Analysis of the likely characteristics of voltage dips *International Scientific-Technical Conference Cybernetics of Energy Systems* **1** 47-52

IOP Conf. Series: Materials Science and Engineering **837** (2020) 012006 doi:10.1088/1757-899X/837/1/012006

Comparative characteristics of the fuel and energy balances of Algeria and Russia and directions for improving their structure

V N Pulyaeva, N A Kharitonova and A Messaoudene

Financial University under the Government of the Russian Federation, 49, Leningradsky Prospekt, Moscow, 125993, Russia

E-mail: kharitonovana1951@gmail.com

Abstract. This study discovers the problem of stable and reliable functioning of the Algerian and Russian energy and fuel complexes. The analysis of structural changing in the composition of the energy and fuel balances of the countries under consideration allowed the authors of the article to conclude that there is a need for a wider use of secondary energy resources, which requires government support measures aimed at attracting investments and switching to providing own secondary energy resources. The possibilities of Algeria in the development of solar and wind energy are shown. For Russia, the relevance of the wider use of biomass, in particular, wood processing waste, was noted. At the end of the work, the conditions are defined, compliance with which will allow the governments of both countries to carry out the necessary transformation of national fuel and energy complexes.

1. Introduction

One of the directions of the economic policy of any state is to ensure the energy efficiency of the national economy, which is determined, inter alia, by the structure of energy and fuel balance (hereinafter - the energy and fuel budget) of the country. The article discusses the features of the structure of fuel and energy and the directions of its improvement for Russia and Algeria. Despite the fact that countries are not comparable in climate and in the industry structure of the national economy, they are united by one thing - a high share of oil and gas revenues in the state budget [1].

For Russia, which occupies the 2nd place in the world in production, as well as in the average daily export of oil and natural gas, oil and gas revenues in the federal budget fluctuate at a level of 40-50% depending on the dynamics of world prices for carbon resources. For Algeria, which is part of the first 20 countries in oil production and in the top ten in terms of its daily exports, and gas in the top ten countries, the share of oil and gas revenues often reaches 60% of the national budget. Before the development of shale oil production, the United States was Algeria's main partner in export deliveries of "black gold". Algeria currently has six oil refineries with a total capacity of 618 thousand barrels per day, which allows exporting products with a higher share of benefit. Over a quarter of the gas exported by Algeria is delivered to consumers in a liquefied state (fifth position in the world ranking) [1-2].

Since 2014, Russia and Algeria have been cooperating in the field of energy. At the same time, a course has been taken to expand bilateral cooperation in the electric power and gas and oil sectors, as well as in the field of renewable energy sources. In addition, it is planned to implement mutually beneficial projects in the field of nuclear energy in Algeria.

2. Materials and methods

A retrospective analysis of the evolution of the energy and fuel complex of Russia, taking into account the dynamics of production, the presence of current and forecast reserves of combustible minerals, as well as trends in world energy, indicates the emergence of a new paradigm for the functioning of the country's energy and fuel complex, the basis of which is the balanced development of the complex until 2040. At the same time, it is assumed not only the extraction of all types of energy carriers, but also their deeper processing, while maintaining the dominant role of gas. For Algeria, which has exclusively oil and gas reserves of energy resources, the immediate prospects for the development of the country's fuel and energy sector are mainly associated with the mining of gas and oil, as well as their deeper processing.

The fuel and energy balance as an instrument of strategic planning for the development of industries and complexes, links production and consumption of various types of energy resources based on a single methodology, generally accepted indicators, according to the approved units of measurement and generally accepted classifications. Fuel and energy complex is developed both in physical units (t, m, kW/h), and in conventional units, which allows using the information contained in it for the purposes of state and municipal administration.

3. Results

The framework of the energy and fuel complex reflects the contribution of each of the types of energy resources to the country's economy, determining to a large extent the directions of its development and the production efficiency of certain areas of industrial and economic activity. It is generally accepted that natural gas is the most economical type of boiler and stove fuel, since it is 2.5–3 times cheaper than coal, 35–40% cheaper than black oil, and 3.5–4 times cheaper than shale oil [4]. Currently, the structure of the fuel and energy complex of the countries under consideration differs significantly from the global level (table 1) [3], which suggests the development of the feasibility of mutually beneficial cooperation in the field of the use of RES.

Table 1. Forecast of the structure of fuel and energy complex for 2020, in percents

Primary resources	World	Russia	Algeria
Coal	30	15	-
Oil	26	22	46
Gas	27	52	54
Atom	3	6	-
Hydro	4	4	-
Bio	4	1	-
Other types of renewable energy sources	6	1	

According to forecasts, the total energy consumption around the world will increase by almost 40% by 2050, will be substantially transformed into the structure of the world energy balance in favor of non-carbon sources (figure 1) [3], which will naturally affect the export opportunities of the countries under consideration.

A promising direction in the development of the global energy market is its transformation, which implies a steady increase in the share of renewable energy sources (hereinafter RES). The main reason for the increased attention to the development of non-carbon energy is due to the need to reduce CO_2 emissions, which should prevent further serious climate change on planet. It should be emphasized that decarburization of the fuel and energy sector requires urgent action on a global scale, which should help reduce carbon emissions and mitigate the effects of climate change. At the same time, energy transformation can be ensured not only thanks to technical and technological developments, the use of modern information and intellectual technologies, but also the wider development of the state policy of supporting the use of renewable energy sources through the use of various market tools.

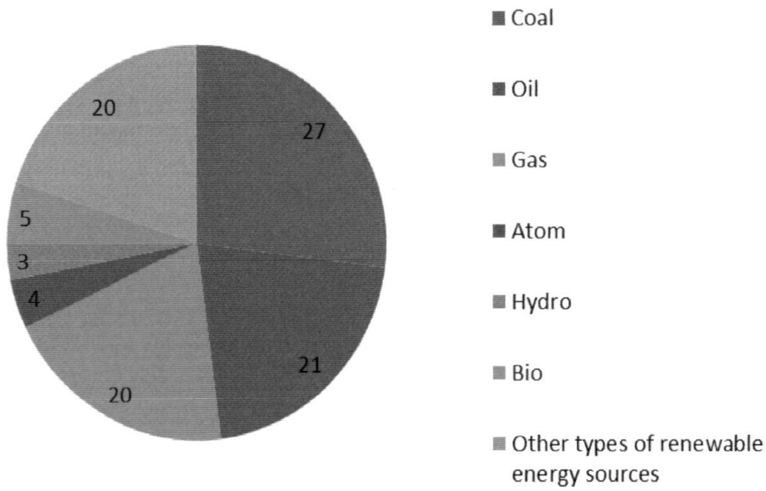

Figure 1. Predicted structure of the global fuel and energy sector by 2050, in percents

Experts from the International Renewable Energy Agency (IRENA) analyzed possible ways of decarburization, and in their study identified the potential for countries, regions and the world as a whole to expand the use of renewable energy sources with an increase in their share in the global fuel and energy sector. It is RES that will satisfy almost 30% of the demand for electricity in 2023, compared with 24% in 2017. It should be emphasized that at present, the use of renewable energy resources is associated mainly with hydropower. At the same time, in the near future (5-10 years) RES will account for more than 70% of the global increase in electricity production. The expected growth will be achieved due to the development of solar, wind energy, as well as bioenergy.

According to the International Renewable Energy Agency, by 2050, electricity could become the most important energy carrier, with a share in the structure of final consumption of energy resources of about 50% (against 20% at present).

Experts suggest that global demand for electricity will be met in the near future due to hydropower at 16% of wind energy - at 6%, solar energy - at 4%, and bioenergy - at 3%.

4. Discussion

Thus, the dynamic development of Russia and Algeria in the near future will be largely determined by the desire of most countries importing energy resources to diversify the structure of their energy market, make more wide use of non-carbon energy sources, focusing on local, renewable energy sources, which will certainly slow down growth hydrocarbon imports, narrowing market niches for both countries under consideration. The transformation taking place in the world energy industry presupposes not only increased competition between producers of fuel and energy resources, but also the demand for fundamentally new technologies in industrial production. As a result, the nature of energy relations between different states will change. Already, most countries are striving for energy self-sufficiency through the use of the latest technological advances, which affects their energy independence and security [4-5].

As for the exporting countries of primary fuel and energy resources, they are moving from a policy of reducing their own production and increasing imports to the maximum use of energy resources in the domestic market, achieving deeper processing of hydrocarbons. For Russia and Algeria, the ongoing energy transformation is associated not only with deeper processing of energy resources, but also with the development of the renewable energy sector.

Emphasizing the huge potential of Algeria in the field of solar energy, it is advisable to note that the country needs a new policy that can ensure the transition from the use of fossil fuels to renewable energy sources as part of the country's sustainable economic development strategy. Despite the fact that currently in the total amount of installed energy capacity, the share of solar energy is less than 1%, the state program involves increasing the capacity of power of solar-powered plants more than twice in the next 5 years. At the same time, a pilot project is being implemented within the framework of the state wind energy program, according to which it is planned to commission 639 MWt of corresponding capacities by 2023.

To stimulate the development of the renewable energy sector in Russia, the Government of the Russian Federation has developed a set of regulatory documents aimed at creating a new energy paradigm in the country. For the first time, a support mechanism for the renewable energy sector was launched in the country in 2011 through the legalization of the installed capacity market, which allows generating facilities operating on renewable energy sources to fully participate in it. However, according to the authors, the predicted value of the renewable energy share indicator, equal to 4.5% in 2024, will be provided mainly due to the more accelerated development of small hydropower.

Among renewable energy sources, a special place is occupied by biomass obtained from wood processing wastes, which is a CO_2-neutral fuel, since biomass emits CO_2 equivalent to that absorbed by the plant during its growth. For Algeria, the issue of reforestation is currently relevant, which is being implemented in accordance with the national reforestation plan, which will allow to increase the production of wood products by 2025, which will require a modern approach to the disposal of the resulting waste.

It should be emphasized that Russia has enormous forest reserves - almost a quarter of the world's resources, which allowed the development of the forestry, pulp and paper and processing industries, which play an important role not only in the economy of the Russian Federation, but other countries with which Russia sells timber. However, existing technologies for wood harvesting and processing are characterized by the formation of a large amount of waste (table 2) [3-4], which are divided into several categories (slab, ends and trimmings, residues of finished materials during the repair of buildings, wood chips and wood dust) depending on their origin or condition.

Table 2. Indicators of waste generation of forestry and woodworking industry.

Type of production	The share of output, in percents		
	Finished product	Waste	Losses
Forestry	63-80	20-37	-
Wood processing	45-55	38-48	7
Wood board production	85-90	5-10	5
Plywood production	40-50	42-52	8
Wood chemical production	62-68	35-38	-

The Russian government has identified the regions where municipal facilities for the production of steam and hot water working on wood waste will be built for the first time. In particular, in the Moscow region it is planned to build such boiler houses with a total capacity of about 280 MWt, and in the Republic of Tatarstan - 55 MWt. However, a more modern approach to the disposal of wood waste is the production of charcoal and pellets. Currently, Russia exports about 1 million tons of wood waste pellets, which are currently shipped mainly to EU countries. With regard to the further development of mutually beneficial cooperation with Algeria, it is of interest to expand trade relations in the supply of charcoal from Russia, as well as pellets, which will make it possible to introduce adjustments to the country's fuel and energy complex in favor of using biomass [6-10].

The peculiarity and complexity of state regulation in the field of RES development is the need to model market mechanisms through legal norms designed to provide a result that is minimally different from what the market could achieve without government intervention. In addition, states need to find the best options for attracting private investment.

5. Conclusion

Thus, current trends in the formation of fuel and energy indicators, as an instrument of industry and territorial planning and management, include:

- An increase in the share of renewable energy sources (sun, wind) in meeting the energy needs of society (for Algeria and Russia);
- Development of new environmentally friendly technologies for the use of coal (for Russia);
- Increasing the production of energy products with a high share of added value (for Russia and Algeria).

The tasks set can be successfully solved through the implementation of a number of specific measures, as well as the mechanism of state energy policy, inextricably linked with the industrial policy implemented by the government of the countries in question with a view to the sustainable development of their economy:

- Creation of conditions ensuring effective competition of interchangeable energy carriers (gas, coal, renewable energy sources);
- Direct state support for the development of nuclear energy;
- Stimulating the sustainable growth of the coal-chemical and petrochemical industries;
- Creating conditions for attracting investments for the development of bioenergy and wider use of renewable energy sources.

References

[1] I A Ivanova et al 2019 *J. Phys.: Conf. Ser.* **1399** 033038
[2] D I Zimnukhova et al 2019 *J. Phys.: Conf. Ser.* **1399** 033097
[3] *Russian statistical yearbook* 2018 (Moscow: Rosstat) 694
[4] *Ministry of Energy of the Russian Federation* Retrieved from: https://minenergo.gov.ru
[5] Sadriddinov M I et al 2020 *IOP Conf. Ser.: Mater. Sci. Eng.* **734** 012051
[6] Zasko V N, Dontsova O I, Osokina I V, Bazhaev M M and Komarova V V 2019 The economic security of the transport industry: model of financial stability improvement *International journal of Management and business research* **9(1)** 136-41
[7] *Energy Strategy of Russia until 2035* Retrieved from: https://minenergo.gov.ru/node/1920
[8] Repnikova V M, Bykova O N, Skryabin O O, Morkovkin D E and Novak L V 2019 Strategic aspects of innovative development of entrepreneurial entities in modern conditions *International Journal of Engineering and Advanced Technology* **8(4)** 32-5
[9] Morkovkin D E et al 2019 *J. Phys.: Conf. Ser.* **1399** 033042
[10] Lopatkin D S et al 2019 *J. Phys.: Conf. Ser.* **1399** 033061

IOP Conf. Series: Materials Science and Engineering **837** (2020) 012007 doi:10.1088/1757-899X/837/1/012007

Assessment of the level of stability and reliability of the electric power complex

A A Gibadullin

State University of Management, 99, Ryazan Avenue, Moscow, 109542, Russian Federation

E-mail: 11117899@mail.ru

Abstract. This study is aimed at assessing the level of reliability and uninterrupted power supply to consumers. The concept of the reliability of power supply to consumers, which should contain the properties of persistence, uninterrupted operation, durability and maintainability, was presented. In addition, the study analyzed reliability indicators, as a result of which it was revealed that the number of accidents is reduced, however, qualitative indicators, including the number of disconnected consumers, average power outages of consumers and the time of outages, are increasing. At the end of the study, the authors proposed a causal relationship model that affects the reliability and uninterrupted power supply to consumers.

1. Introduction

The electric power complex of Russia is one of the largest in the world, which is combined into a single energy system, including power plants, electric systems and networks, substations, switchgears and elements responsible for the supply of energy to the final consumer. The complex took a rather long period, for example, in 1920, the State Plan for Electrification of Russia was adopted and then, until 1941, the industry developed stable indicators, even exceeding the tasks set. In the 50s of the 20th century, the potential of the Soviet electric power industry was created, and in the 60-70s, the existing power plants were built and put into operation, an integrated power system was created and the basic principles of the functioning of the electric power complex were formed.

During the period of Russia's transition to market relations, the electric power complex also underwent a transformation, as a result of which competitive and monopolistic activities were singled out and new companies were formed in the industry that are privately owned and independently responsible for the operation of electric power facilities. At the same time, during the transition to innovative and digital technologies, the work of highly efficient production and various types of activities that ensure human life, issues of reliability and uninterrupted power supply to consumers come to the fore, which requires the complex to provide the required amount of requested load with electricity [1].

The main construction of electric power facilities producing electric energy today was carried out about half a century ago. Despite the fact that developed countries have already repeatedly switched from one type of production to another, the electric stations created in the Soviet years are operated in Russia. This situation causes a number of negative consequences for both industry and consumers, for example, equipment has a high level of wear, a high share of energy consumption, low efficiency and installed capacity utilization, low innovative and technological development of electric power facilities and others Problems. Thus, these negative factors can adversely affect the reliability and uninterrupted power supply, which, as a result, can violate not only the stability of the entire electric power system in Russia, but also exacerbate the internal and external situations of the national economy.

2. Materials and methods

The main goal of the work is to assess the reliability and uninterrupted power supply to consumers. The following tasks were formed in the work:

- determine the property of reliability and present its elements;
- present a causal relationship model that affects the reliability and uninterrupted power supply.

The work used data from line ministries and state statistical agencies, and the scientific methods used allowed us to solve the research tasks.

3. Results

The reliability of the electric power complex is an essential component of the national economy, the violation of which can lead to adverse consequences for the types of economic activity. The reliability property meets a number of characteristics, which includes the following main components:

- persistence - this is the property of an object to function for a long period of time, while not losing the assigned properties and characteristics in time;
- uninterrupted operation - this is the property of an object to continuously operate at a given operating mode of equipment;
- maintainability - this is a property of an object that allows you to restore or technically change individual elements or units of equipment in order to maintain the future operation of the equipment;
- longevity is a property of an object that ensures continuous operation of the equipment from the beginning of its commissioning to the end of the equipment's park life.

Thus, the reliability of power supply is a complex property that includes many elements, the preservation of which ensures the stability of the functioning and development of the electric power complex. Of course, a change in these properties negatively affects the entire system, and as a result, disrupts the power supply to consumers, which can lead to production stoppage, mass rejection of products, worsening of the living conditions of the population, changes in the operation of individual city systems, etc.

However, despite the fact that reliability is a complex technical system, and maintaining the integrity of this system ensures the stability of power supply and the entire energy system, there are problems associated with ensuring the reliability of power supply to consumers.

For an objective analysis, we consider the indicators characterizing the cessation of power supply to consumers (figure 1) [2-3].

From the presented figure it can be seen that the number of mass power outages during the period under review is generally decreasing, however, despite this, quality indicators are increasing. So, for example, from 2017/18 to 2018/19, the number of power outages decreased from 37 to 19 cases, and the average number of disconnected consumers increased from 16.920 thousand to 24.070 thousand people. Thus, a decrease in the number of outages is not an indicator of an increase in the reliability and uninterrupted power supply to consumers.

Next, we consider the average power outages of consumers and the duration of outages (figure 2) [2-3].

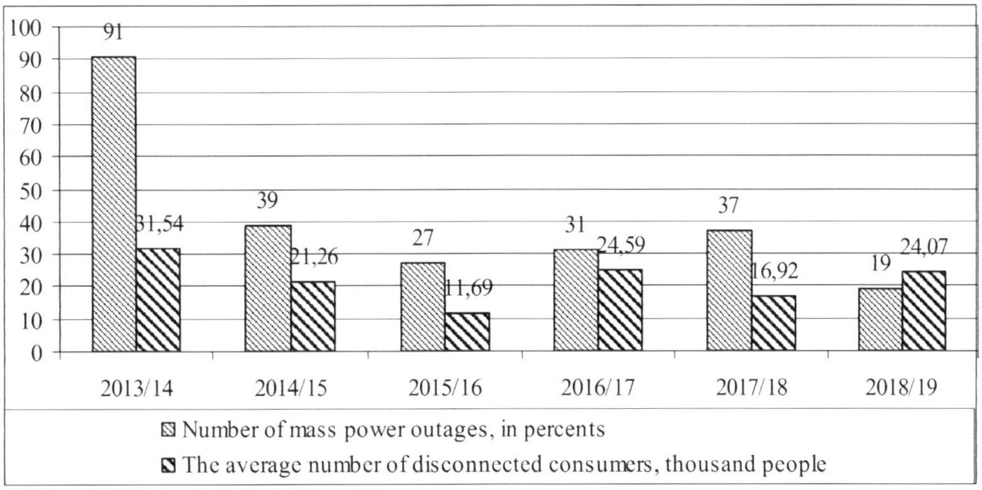

Figure 1. Cessation of energy supply to consumers in the heating season (from November to March).

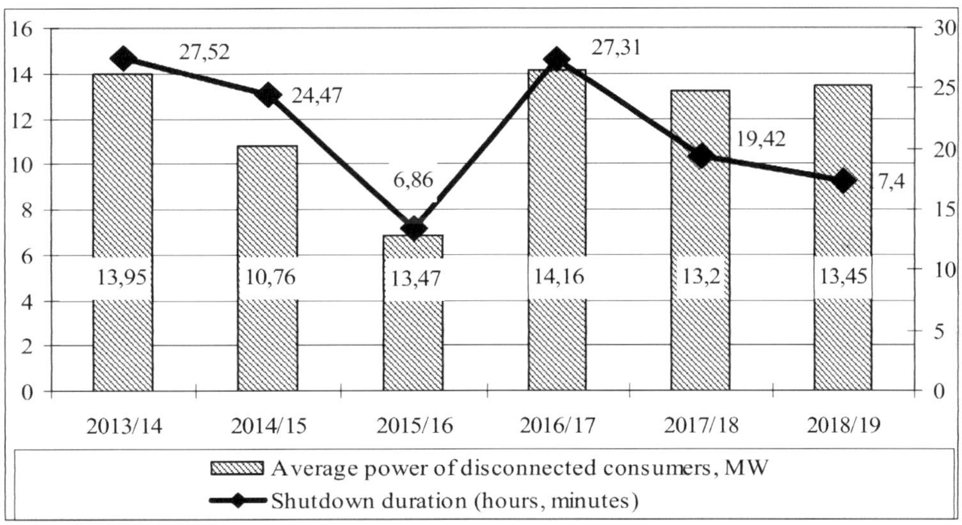

Figure 2. Average power of disconnected consumers (MW) and duration of outages.

The figure shows that the duration of the outages has a changing curve, which, in general, decreases during the period under review, and the average power outage of consumers, although it has decreased, is insignificant.

In order to ensure reliable and uninterrupted power supply at the state and corporate level, measures are being taken to preserve the power supply to consumers, which are associated with updating and modernization of production facilities, carrying out diagnostic and repair measures, restoration and technical impact on electric power facilities. At the same time, not all of these factors are components in violation of the reliability of power supply to consumers, in this regard, the formation of a causal relationship model in which to present factors affecting the provision of uninterrupted power supply to consumers is relevant [4-5].

4. Discussion

We will form a model of causal relationship affecting the provision of reliable and uninterrupted power supply to consumers (figure 3).

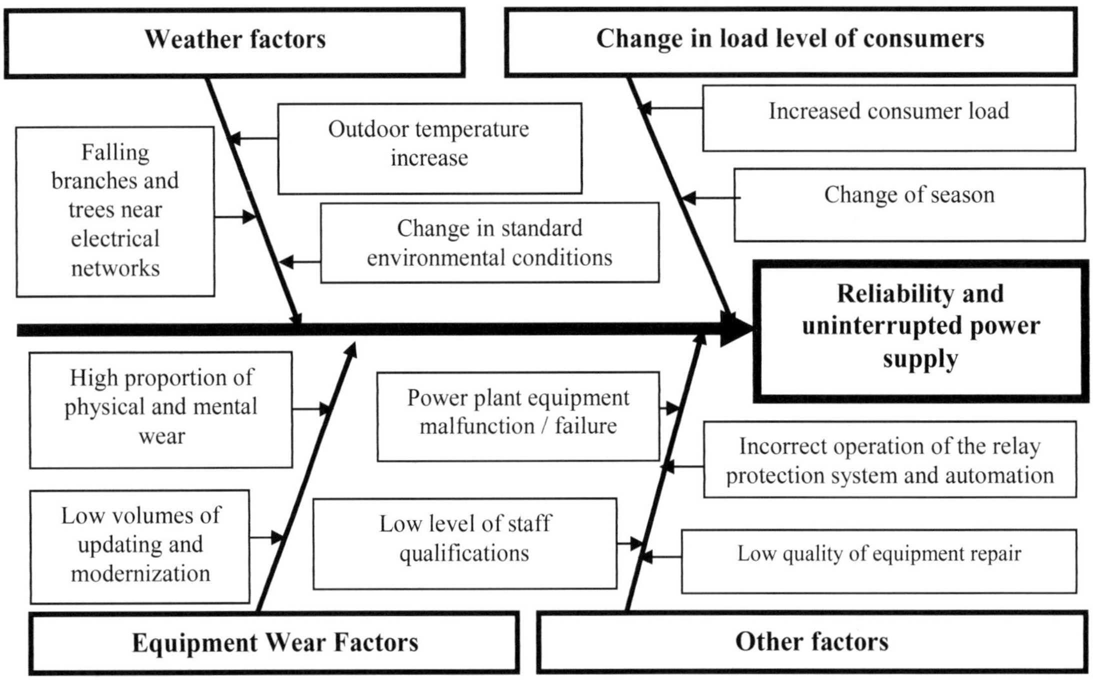

Figure 3. A causal relationship model that affects the reliability and uninterrupted power supply to consumers.

The presented model contains weather factors, changes in consumer load level, equipment wear factors and other factors affecting the reliability and uninterrupted power supply.

Today, activities in the field of improving the reliability and uninterrupted power supply are taking place in certain areas of activity, for example, electric networks monitor the glades, and the System Operator of the UES of Russia monitors the load level and, if necessary, can use non-operated production capacities. However, in the opinion of the researchers, the problems associated with improving the reliability and uninterrupted power supply must be solved systematically, namely, to ensure the management of the factors presented in the model and to carry out activities aimed at preventing the development of negative factors affecting the electric power complex.

5. Conclusion

The study was devoted to assessing the reliability and uninterrupted power supply to consumers. In the framework of this work, it was revealed that reliability has a number of properties - it is the maintainability, uninterrupted operation, durability and maintainability of the system, the interconnection of which ensures the safe and stable operation of the energy system. In the work, indicators were analyzed that affect the reliability of power supply, as a result of which it was revealed that the number of accidents has been decreasing in recent years, but their negative consequences are increasing. At the end of the study, a model was proposed to ensure the reliability and uninterrupted power supply, which should be considered as a single system and to carry out holistic management of negative factors.

IOP Conf. Series: Materials Science and Engineering **837** (2020) 012007 doi:10.1088/1757-899X/837/1/012007

References

[1] Gibadullin A A et al 2020 IOP Conf. Ser.: Earth Environ. Sci. **421** 032051
[2] *Russian statistical yearbook* 2018 (Moscow: Rosstat) 694
[3] *Ministry of Energy of the Russian Federation* Retrieved from: https://minenergo.gov.ru
[4] *Energy Strategy of Russia until 2035* Retrieved from: https://minenergo.gov.ru/node/1920
[5] Rudaya I L 2019 Issues of strategic management of Russian innovative organizations' intellectual property. *Upravlenie* **7** 1 50-9

IOP Conf. Series: Materials Science and Engineering **837** (2020) 012008 doi:10.1088/1757-899X/837/1/012008

Renewable energy engineering solutions

J N Hojiev[1], D E Morkovkin[2], V G Starovoitov[2], A A Gibadullin[3], A V Dubrovsky[2] and I A Boldyreva[4]

[1] The Tajik state university of finance and economics, 64/14, Nahimova Street, Dushanbe, 734055, The Republic of Tajikistan
[2] Financial University under the Government of the Russian Federation, 49, Leningradsky avenue, Moscow, 125993, Russian Federation
[3] State University of Management, 99, Ryazan Avenue, Moscow, 109542, Russian Federation
[4] Novocherkassk Engineering Institute reclamation Don State Agrarian University, 111, Pushkinskaya Street, Novocherkassk, 346428, Russian Federation

E-mail: 11117899@mail.ru

Abstract. As part of the study, engineering solutions in the field of creating power plants based on renewable energy sources were analyzed. The work revealed that in Russia the level of development of renewable energy sources is ten times lower than in the world, and renewable energy potential is not used. An engineering solution was proposed in the work, which consists of the stages of analysis, design, construction and commissioning of a power plant based on renewable energy sources. At the end of the study, a scheme was proposed for introducing wind power plants into the energy system in order to maintain the stability and reliability of power supply to consumers and meet the demand for electric energy.

1. Introduction

One of the main advantages of modern electric power industry is the possibility of using clean technologies, which reflect the level of efficiency and progressiveness of the national electric power industry [1-3]. Number of energy complexes have already abandoned the use of polluting and inefficient technologies several times and switched to new technologies for generating electric energy [4].

In the global energy sector today, the issues of transition to environmentally friendly and safe technologies are being actively discussed; in this regard, various engineering solutions are being developed to ensure the transition of national energy companies to renewable energy sources. In Russia, the issues of switching to renewable energy sources were laid bare back in the 2000-s. At the same time, for example, a completely different situation is observed in European countries, where about 10% of all capacities operate based on renewable energy sources and at the same time, there is a rejection of traditional power plants [5].

The heterogeneity in the development of renewable energy sources on the territory of the world justifies this study in order to identify the possible development of renewable energy sources and to search for the most optimal engineering solutions to create a new electric power potential of the Russian electric power industry.

IOP Conf. Series: Materials Science and Engineering **837** (2020) 012008 doi:10.1088/1757-899X/837/1/012008

2. Materials and methods

The aim of the study is to assess the potential of renewable energy sources in Russia and to develop an engineering solution aimed at the development of renewable energy. The following tasks were formed:

- Assess the level of use of renewable energy;
- Development of engineering solutions in the field of energy.

Scientific research was based on an assessment of engineering and design decisions of various organizations in the field of energy. In the work, various scientific approaches and methods were used.

3. Results

Currently, energy development is uneven, individual states have come forward in terms of these indicators, which reach 20-30% of the total installed capacity, other states only adopt directives for the development of renewable energy. The power industry of Russia developed during the period of the USSR, and today in all the republics that are members of the Soviet Union, the same problems have been revealed, among which are the high level of moral and physical deterioration of equipment, a decrease in efficiency and resource saving in the industry, a high proportion the use of obsolete technologies, a low share of the introduction of innovative and advanced technologies in the industry, a high share of the environmental load on the environment and other problems. Can solve problems by modernization and renewal of the capacity of power plants or the creation of a fleet of electrical installations operating on renewable energy sources [6-8].

Assess the use of renewable resources for the electricity industry (figure 1) [9].

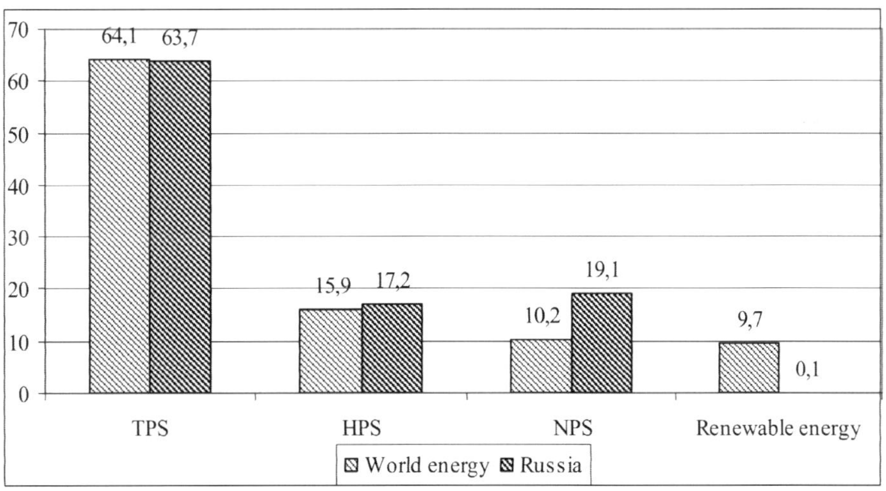

Figure 1. The share of energy generation at different types of stations.

Figure click that the level of development of the world renewable energy is several times higher than the indicators of the Russian electric power industry, which is associated with the presence of excess capacities and low level of use of renewable energy sources.

Consider the potential of renewable energy in the European part (figure 2) [10].

From the presented figure it can be seen that the number of sunny and clear territories in the European part is not more than 20%, and the wind force reached 18 meters per second is observed in a separate area, while the average wind speed in the European part is 9 meters in give me a sec [11-13].

It should be noted that it is advisable to install wind farms in territories where the wind speed exceeds 18 meters per second, which will allow the most efficient use of production capacities. It is advisable to place solar power plants in areas where the number of sunny days reaches maximum values [14].

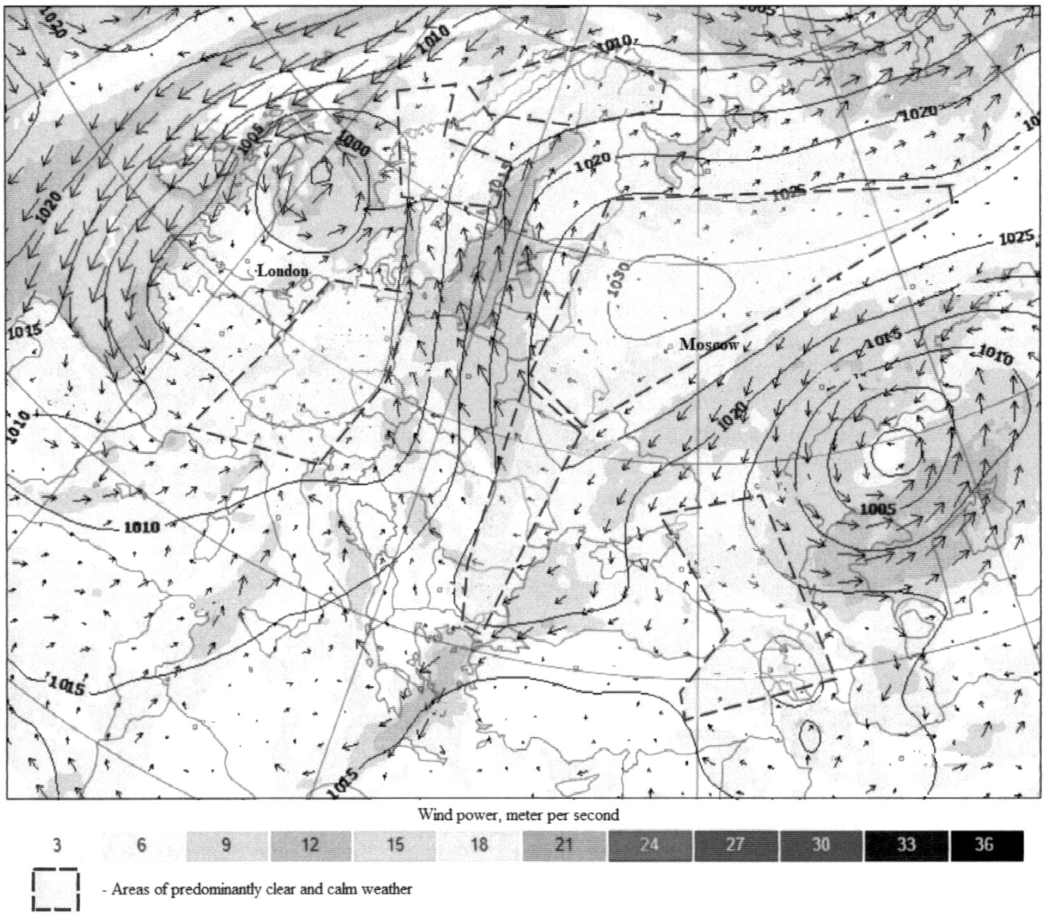

Figure 2. Renewable energy potential.

Thus, in Russia this type of energy is not developed and requires the development of engineering solutions to increase the importance of using renewable energy.

4. Discussion

Engineering decisions in the field of development of renewable energy sources for the Russian Federation should be based on the following special principles [15-18]:

- Assessment of the level of reliability, stability and uninterrupted operation of the current generation;
- Analysis of the level of depreciation, renewal and retirement of production facilities, the volume of planned projects to upgrade energy equipment;
- Search for the most vulnerable and emergency areas where frequent power outages occur;
- Decision-making on the construction of a station based on renewable energy sources;
- Assessment of capital and operating costs;
- Development of design solutions in the field of station construction;
- Carrying out the construction of a power plant;

IOP Conf. Series: Materials Science and Engineering **837** (2020) 012008 doi:10.1088/1757-899X/837/1/012008

- Commissioning and maintenance.

Thus, the engineering solution for the construction of power plants based on renewable energy sources for the Russian Federation should contain a similar principle for the construction of project activities, in which all the necessary measures are carried out from the substantiation of the decision to the commissioning of the constructed station.

However, that such projects in the field of construction from power plants based on renewable energy sources will ensure the stability, reliability and safety of the entire electric power complex. It is possible to achieve such stability in the prism of the energy system due to the convergence of power plants based on renewable energy sources into a common electricity system (figure 3).

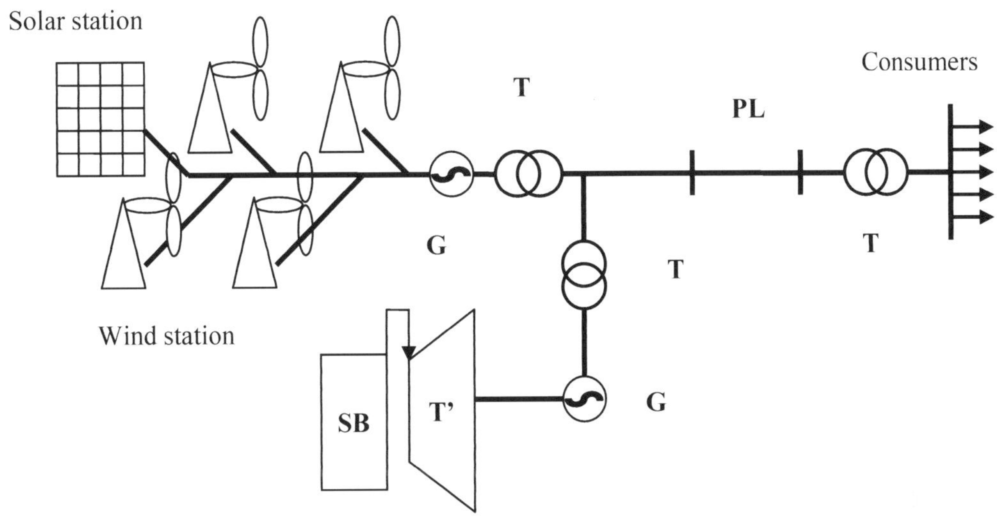

Figure 3. The system of interconnection of renewable energy sources with the energy system.

Thus, in order to maintain the power system in a stable state, it is necessary to converge power plants operating on the basis of renewable energy sources into a single system in order to achieve a balance in the production and consumption of electric energy, both in the normal mode and in the mode of disruption of the power supply system.

5. Conclusion

The analysis in the field of renewable energy development showed that the design and construction of such power plants should be based on taking into account the amount of the missing system load, assessing the level of reliability, uninterrupted operation and stability of electric power facilities and depreciation of production capacities. However, the study suggested that renewable energy facilities should be included in the Unified Energy System to ensure the sustainability and reliability of the entire electricity industry.

References

[1] Davnis V V, Tinyakova V I, Blinov A O and Volodin Yu V 2019 Combined Modeling of Projected Evaluation of the Regional Socio-economic Development *International Journal of Economics & Business Administration (IJEBA)* **0** 1 348-54

[2] Kobtseva O N, Novoselova N N, Novoselov S N, Shichiyakh R A, Morkovkin D E and Sidorchukova E V 2017 Organizational and economic features of import substitution formation and realization in the conditions of spatial restrictions *International Journal of Applied Business and Economic Research* **15(23)** 25-35

[3] Ivanova I A et al 2020 *IOP Conf. Ser.: Earth Environ. Sci.* **421** 032039
[4] Yakunina G E 2019 Research of digital communications models within organizations and at the state level in the countries-leaders in the use of digital communication technologies *E-management* **2** 4 41-50
[5] Tolkachev S A et al 2020 *IOP Conf. Ser.: Earth Environ. Sci.* **421** 032041
[6] Zakharov V N, Linnik V Y, Linnik Y N and Zhabin A B 2019 Classification of coal seams by features of geological structure and characteristics of breaking *Mining Informational and Analytical Bulletin* **5** 5-12
[7] Bryukhovetskaya S V et al 2020 *IOP Conf. Ser.: Earth Environ. Sci.* **421** 042018
[8] Sadriddinov M I et al 2020 *IOP Conf. Ser.: Mater. Sci. Eng.* **734** 012051
[9] *Russian statistical yearbook* 2018 (Moscow: Rosstat) 694
[10] *Ministry of Energy of the Russian Federation* Retrieved from: https://minenergo.gov.ru
[11] Ustyuzhanina E, Evsukov S and Komarova I 2018 Network economy as a new economic system *European Research Studies Journal* **21(3)** 77-89
[12] Romanova Ir N et al 2020 *IOP Conf. Ser.: Mater. Sci. Eng.* **734** 012166
[13] Sozontov A, Ivanova M and Gibadullin A 2019 Implementation of artificial intelligence in the electric power industry *E3S Web of Conferences* **114** 01009
[14] Gibadullin A A et al 2020 *IOP Conf. Ser.: Mater. Sci. Eng.* **734** 012170
[15] Linnik Y N, Linnik V Y, Zhabin A B and Polyakov A V 2019 Integrated estimation of strength properties of complex-structure coal seams *Mining Informational and Analytical Bulletin* **8** 33-41
[16] Gibadullin A A et al 2020 *IOP Conf. Ser.: Earth Environ. Sci.* **421** 032051
[17] Zakharov V N, Linnik V Y, Linnik Y N and Zhabin A B 2019 Classification of coal seams by features of geological structure and characteristics of breaking *Mining Informational and Analytical Bulletin* **5** 5-12
[18] Morkovkin D, Lopatkin D, Sadriddinov M, Shushunova T, Gibadullin A and Golikova O 2020 Assessment of innovation activity in the countries of the world *E3S Web of Conferences* **157** 04015

IOP Conf. Series: Materials Science and Engineering **837** (2020) 012009 doi:10.1088/1757-899X/837/1/012009

Assessment of properties of coal seams as destructible environments

V Yu Linnik[1], Yu N Linnik[1], A Zich[2], S V Makar[3], Ch V Kerimova[3] and M I Sadriddinov[4]

[1] State University of Management, 99, Ryazan Avenue, Moscow, 109542, Russian Federation
[2] MSQF GmbH, 111, Dorfstrasse, Oderwitz, 02791, Germany
[3] Financial University under the Government of the Russian Federation, 49, Leningradsky avenue, Moscow, 125993, Russian Federation
[4] Tajik State Finance and Economics University, 64/14, Nakhimova Street, Dushanbe, 734067, Republic of Tajikistan

E-mail: vy_linnik@guu.ru

Abstract. The article analyzes indexes that characterize properties of coal seams and are used in calculation of loads on cutting tools of mining machines, or in selection of drives. Determination of geological structure complexity of coal seams takes into account hard inclusions and dirt bands in coal, as well as their size and lithology type. The notion of coal cuttability in spalled and unspalled face zones is explained using the coefficient of coal spalling. For strength estimation of complex structure coal seams with dirt bands and hard inclusions, it is proposed to use the index of integral cuttability which is the sum of coal cuttability and generalized index of densities and properties of inhomogeneities in the seam. In geological documentation, strength characteristics of inhomogeneities are described using the ultimate compression strength or Protodyakonov factor of hardness (in Russia). For this reason, the authors propose an empirical expression to recalculate cuttability of inhomogeneities in terms of uniaxial compression strength or hardness factor. The estimation algorithms of brittleness, grindability and breakability of coal seams are presented. The formula of coal abrasiveness is given. Systematization of data on breakability of coal seams has made a framework for classification of coal seams as media disintegrated by cutting tools of mining machines.

1. Introduction

Even though the share of coal generation in the modern world is gradually decreasing, its share is still quite large and is about 40%. In this regard, issues related to sustainable and uninterrupted coal production are relevant [1-4].

Modern ideas about the properties of coal seams as objects of destruction by cutters of coal-mining machines are based mainly on the following works of Russian scientists [6; 9] and foreign scientists [7; 8; 10-20] considering joint study of coal seams and fracture processes. The obtained parameters of the strength properties of the layers serve as the basis for the calculations of the power requirements of coal mining machines.

Russian researchers use mainly experimental and statistical methods, which distinguish them from the studies of foreign authors that are based on physical representations of the laws of the processes of cutting coal and rocks. As the main indicator that determines the properties of coal seams Russian scientists recommended [5-6; 9; 21; 22] the cuttability of coal seam. To measure this value as an integral strength property depending on the complexity of coal seams structure and geological factors [6; 21; 23] the special equipment and techniques have been created [22; 24]. The complexity of coal seams structure and geological factors includes the abrasiveness of coal seams and the content of solid inclusions in them [25], the ability of coal to grind during cutting, their destructibility and brittle-plastic properties [26; 27].

In general, the above-mentioned studies of Russian scientists allowed classifying coal seams according to their cuttability and geological structure [23].

Among foreign researchers, special attention should be paid to the studies of I. Evans and S. Pomeroy [10; 11], who were among the first to propose models and give a physical interpretation of the laws of the processes of coal destruction during cutting. The first such model [10] was proposed by Evans in relation to the determination of the cutting force. According to this model, when the cutting tool is inserted into a seam, radial compressive stresses occur in it and when they reach the tensile strength of the material, a cleavage occurs in the form of a symmetrical V-shaped fragment. This assumes that there is no friction between the tool and the rock and assumes that the normal contact pressure between the cutter and the rock mass is evenly distributed. Later, Goktan R. improved the theory of cutting Evans applied to cutting rocks cutters with a conical shape of the cutting part [12].

Along with theoretical methods, numerical methods to describe the destruction of rocks with a cutting tool have been developed. For example, in [19] the destruction of rocks by a tangential cutter was studied in two-dimensional and three-dimensional formulations, and in [20] the finite difference method was used in the three-dimensional formulation of the problem using the Particle Flow Code (PFC) algorithm. The authors explained the choice of the discrete element method tool for numerical modeling by the fact that this method allows modeling a rock mass as a collection of particles with different properties, including pores, cracks, cavities, etc. [27]. The study consisted in comparing the results of numerical modeling with laboratory test data and calculated values. It was noted that the results of numerical modeling and theoretical calculations agree well with each other, but poorly with laboratory data.

Paying tribute to the authors of the studies above, it should be noted that the theoretical models of destruction of coal and rocks by the cutters proposed by them are made with multiple assumptions and simplifications. The main of them are the followings: cutting is considered as a two-dimensional process, while it is believed that there is no interaction between adjacent cuts, ignored the presence of solid inclusions and strong rock layers in coal seams, as well as possible fractures in coal seams during their destruction [8; 10; 14; 27].

Summarizing the results of research by scientists from Russia and other countries, it should be noted that in relation to the description of the efficiency of the operation of coal mining machines, it is necessary to determine a set of characteristics that best reflect the process of destruction of the coal seam. At this, it is especially important that such characteristics should affect the failures of cutterheads. Ultimately, it is necessary to determine and justify a generalized criterion for the properties of destructible coal seams, which would comprehensively evaluate layers of complex structure as environments destroyed by cutterheads of coal mining machines.

For efficient coal mining as well as mining machinery design and operation coal seams should be assessed by taking jointly:

- Structural characteristics;
- Breakability indexes (cuttability and brittleness).

Structural heterogeneities of coal seams have influence on loading and reliability of mining machines and should primarily be considered when selecting parameters of cutting tools and transmissions.

Properties of a coal mass are variable, which is mainly caused by in homogeneities (dirt bands and hard inclusions) and by rock pressure effects. For this reason, physical magnitudes turn out to be inadequate for the description of a coal seam as a destruction object. It is advisable to use process variables of coal seam breakability as they allow integral assessment of coal seams as media destroyed by cutting tools of mining machines.

In Russia engineering designs use such characteristics as: cuttability of coal, A_{coal}, N/mm, and coal seam, A_{seam}, N/mm, hard inclusions A_{hard}, N/mm, and dirt bands, A_{dirt}, N/mm, densities of hard inclusions, S^{*}_{hard}, %, and dirt bands, S^{*}_{dirt}, %; coal seam abrasiveness ρ (mg/km); coal brittleness B and grindability m, which are dimensionless values.

2. Materials and methods

The aim of the study is to evaluate the properties of coal seams as destructive media. The following tasks are set in the work:

- Assess the complex geological structure of coal seams;
- Assess the fragility and grind ability of coal, the level of destructibility of coal beds;
- Present a classification of coal seams by fragility and geological structure.

The work uses scientific and experimental methods, which made it possible to conduct an experiment and evaluate coal seams as destructive media.

3. Results

Structural complexity of coal seams is governed by dirt bands and inclusions harder than coal. Hard inclusions commensurable in size with cutting depth essentially add to load on cutting tools (especially dynamic load) and worsen operation of mining machines.

Hard inclusions are divided into types with regard to their dimension as per data in table 1.

Table 1. Data for plotting of the dependence.

Type of inclusion	Length, cm	Thickness, cm	Cross section, cm^2
Fractured	To 0	To 5	To 50
Unfractured	10–30	5–10	50–300
Consolidated	Larger than 30	More than 10	Larger than 300

Lithology types of hard inclusions are diverse. Most often there are carbonate-bearing hard inclusions divided into 4 groups of petrography: carbonate (calcite) K1; carbonate (siderite) K2; carbonate K3; carbonate K4. In coal geology, the carbonate inclusions are mostly named as pyrites. The hard carbonate inclusions widely range in size. However, their length (from 40 to 315 cm.) generally essentially exceeds their thickness (from 15.5 to 45 cm.).

The carbonate inclusions have various strength characteristics. Their ultimate compression strength σ_{com} ranges from 51.0 to 74.7 MPa and is on average 65.3 MPa. Pyrite inclusions show the highest strength (σ_{com} = 107.4–117.6 MPa), although their small size, according to practice, has minor influence on efficiency of coal mining machines. Conversely, large and numerous consolidated hard inclusions, if present, greatly complicate operation conditions of cutting tools, which limits capacity of the machines in the long run.

In calculations connected with determination of load on cutting tools or with choice of electric drives, hard inclusions or dirt bands are estimated by the index of their density in coal seams. The density of inclusions (dirt bands) is determined by their specific area found from the formula:

$$S^{*} = 100\frac{\sum s}{F} \quad (1)$$

where $\sum s$ is the total cross-section area of hard inclusions or dirt bands in the measurement interval along the face, m^2; F is the exposed area of the face, m^2.

Based on the information about the number and types of hard inclusions from the mine geology documentation, their density in a coal seam may approximately be assumed as in table 2.

Table 2. Density of hard inclusions in coal seam.

Number of hard inclusions per 100 m of the face length	Density of hard inclusions per 1 m of coal band thickness, %		
	Fractured	Unfractured	Consolidated
10	0.05	0.3	1.5
20	0.1	0.6	3.0
100	0.5	3.0	15.0

Density of hard inclusions in coal seams having different thickness and number of inclusions per 100 m of face length than the seams in Table 2 is determined by interpolation.

The strength characteristics of dirt bands, similarly to hard inclusions, vary in wide ranges within a single lithology type. Sand bands feature the highest strength (lithology types P4–A1 and A1–P4): their ultimate compression strength σ_{com} = 53.0–101.8 MPa and the Protodyakonov index of hardness f = 5.9–8.0.

The softest rocks are argillite (lithology types Ar_1, Ar_2) and carbonaceous argillite (Ar_c). The ultimate compression of argillites varies from 16.6 to 34.6 MPa and f = 1.5–2.7. Siltstone takes intermediate position between sandstone and argillite although the compression strength also ranges widely: σ_{com} = 23.0–55.5 MPa; A_{dirt} =208–565 N/mm, f = 2.8–5.0.

Density of dirt bands in the assessment of coal breakability is determined similarly to density of hard inclusions. Average thickness of a dirt band per face length is determined together with the number of dirt bands, their lithology types and density of each lithology type (in case there are more than one dirt bands); then, the average cuttable thickness of coal seam is found.

The cuttability is understood as the ability of coal to sustain mechanical attack of a cutter of standard geometry. The cuttability index is assumed as A, which is the increment in the cutting force per 1 cm. depth of cutting by a measurement cutter with a cutting edge 2 cm. wide, equipped with a force gage, in a standardized mode.

$$A = \frac{Z}{h} \quad (2)$$

where A is the cuttability, N/mm; Z is the cutting force, N; h is the cut depth, mm.

In operating mines in Russia, cuttability of coal is determined using drilling dynamometers. The method consists in drilling-out of a pre-drilled hole with a diameter of 42 mm, at a constant drill feed, with record registration of change in the cutting resistance moment.

Depending on the problems to be solved, cuttability of coal is determined in spalled and unspalled zones.

Cuttability of coal in unspalled zone A_{coal} is determined deep in rock mass free from spalling. Owing to this, this is the most stable value and, therefore, is used as the base classification indicator of application areas for coal mining machines.

Cuttability of coal in face area A_B is lower than in unspalled zone due to rock pressure. The zone of rock pressure manifestation is named a spalling zone (as a rule, makes 0.4–0.6 m of coal seam thickness) and a pronounced spalling zone (10–20 cm deep brim of the face). Cutters of mining machine drum operate, partly or totally, within the spalling zone; for this reason, the cuttability of coal for mining machine with the cutting width is found from the expression:

$$A_B = k_{spall} A_{coal} \tag{3}$$

where k_{spall} is the spalling coefficient which is a ratio of coal cuttability in unspalled zone to average cuttability of coal in operation zone of cutting drum.

It is experimentally found that the average coefficient of spalling for a cutting drum with the cutting width B_{cut} is given by:

$$k_{spall} = k_{spall_0} + \frac{(B_{cut}/H_{seam}) - c}{(B_{cut}/H_{seam}) + d} \tag{4}$$

where k_{spall_0} is the coefficient of spalling at the face edge; H_{seam} is the coal seam thickness, m; c and d are the coefficients depending on coal quality and mining conditions. The calculations usually assume: $c = 0.1$ – for a hard coal, 0.15 – for a lignite coal, $d = 1.0$ for $B_{cut} < 1.2$ m, 1.3 for $B_{cut} > 1.2$ m.

Integral cuttability index of coal seam - A_e, N/mm. Coal seams contain usually both dirt bands and hard inclusions. In these cases, it is proposed to assess such complex-structure seams using the integral cuttability index A_e, N/mm.

$$A_e = A_{coal} + A^*_{inh} \tag{5}$$

where A^*_{inh} is the generalized index of density and properties of inhomogeneities (dirt bands, cutbacks, hard inclusions) in coal seam, N/mm.

The value A^*_{inh} characterizes level and frequency of maximum loads arising in cutters and responsible for the cutting process dynamics and cause abrupt failures in destruction of complex-structure coal seams. This index is calculated from:

$$A^*_{inh} = A^*_{hard} + A^*_{dirt} \tag{6}$$

where A^*_{hard} is the generalized index of density and properties of hard inclusions, N/mm; A^*_{dirt} is the generalized index of density and properties of dirt bands, N/mm:

$$A^*_{hard} = 1.5 A_{hard} K_{hard} S^*_{hard} d_{hard}\text{, N/mm} \tag{7}$$

$$A^*_{dirt} = A_{dirt} K_{dirt} S^*_{dirt}\text{, N/mm} \tag{8}$$

where S^*_{hard}, S^*_{dirt} are the densities of hard inclusions and dirt bands in coal seam, respectively, %.

The coefficient $K^*_{inh}(K_{hard}, K_{dirt})$ is given by:

$$K_{inh} = \frac{A^*_{inh} - A_{coal}}{A_{coal}} \tag{9}$$

Ratio of cutback inclusions by cutters d_{hard} is advised to be accepted as 0.87 for fractured inclusions, 0.9 for unfractured inclusions and 0.99 for consolidated inclusions.

It is difficult to show the influence exerted by each of A^*_{hard} and A^*_{dirt} on A^*_{inh}. This is explained by the fact that the weight of each index is comprised of a series of characteristics such as thickness of an inclusion or dirt band, cuttability, lithology type. Put it otherwise, hard inclusions having high cuttability but low density in a coal seam have weaker effect on A^*_{inh} than a thick dirt band of lower cuttability. Physically, the index A^*_{inh} means reduced cuttability of inhomogeneities (with regard to S^*_{hard} and S^*_{dirt}).

The analysis of the expression (5) shows that A^*_{inh} is either always equal to A_{coal} (in coal seams of simple structure), or higher than A_{coal}; in the latter case, the difference grows with the density of inclusions in the coal seam and with the discrepancy between the cuttabilities of coal and inclusions.

One of the main factors to govern the value of A_{coal} and A_e, and, consequently, loads imposed on cutters is the presence and cuttability of hard inhomogeneities in coal seams (hard inclusions, dirt bands and cutback host rocks). Mine geological documents on coal seams characterize strength of inhomogeneities by the ultimate compression strength σ_{com} or by the Protodyakonov index of hardness f (in Russia).

The analysis of the experimental data on strength characteristics of hard inclusions and dirt bands discloses a correlation between them depicted in figure 1.

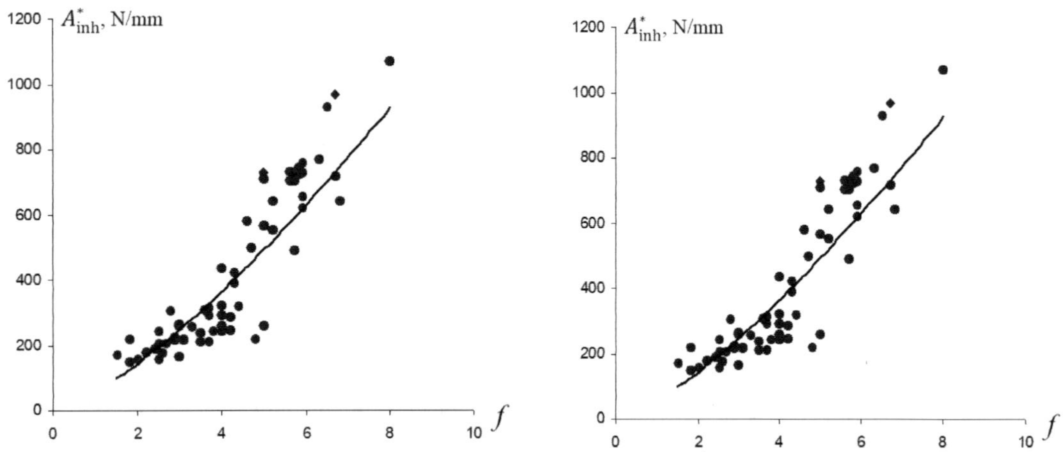

Figure 1. Relationship between cuttability A^*_{inh}, uniaxial compression strength σ_{com} and Protodyakonov index of hardness f of inhomogeneities in coal seam.

The empirical expressions to describe the relationships in Figure 1 are given by:

$$A^*_{inh} = 0.3\sigma^{1.19}_{com} \quad (10)$$

$$A^*_{inh} = 56.3 f^{1.35} \quad (11)$$

Regarding the latter expression (7) and (8), we rewrite the formulas for the generalized index of density and properties of inhomogeneities in coal seam as follows:

- By mentioned σ_{com}:

$$A^*_{inh} = \sigma^{1.19}_{com}(0.45K_{hard}S^*_{hard}d_{hard} + 0.3K_{dirt}S^*_{dirt}) \quad (12)$$

- By mentioned f:

$$A^*_{inh} = f^{1.35}(84.4K_{hard}S^*_{hard}d_{hard} + 56.3K_{dirt}S^*_{dirt}) \tag{13}$$

The obtained relations $A^*_{inh} = f(\sigma_{com})$ and $A^*_{inh} = f(f)$ enable finding A^*_{inh} at the known values of σ_{com} and f, which makes it possible to calculate the index of equivalent cuttability A_e of coal given there is any available geological information on strength characteristics of inhomogeneities, and this essentially simplifies computation.

It is known that for any process connected with cutting of coal and rocks, it is valued that the rate (efficiency) of the process, any other conditions being equal, is proportional to the cutting power P and inversely proportional to the exposed coal area S and specific energy intensity of cutting N_W. For instance, the feed velocity v_{feed} is calculated as:

$$v_{feed} = \frac{P}{60B_{cut}H_{seam}H_W}, \text{ m/min} \tag{14}$$

where P is the cutting power, kW; B_{cut} is the cut width, m; H_{seam} is the cuttable thickness of coal seam, m.

Thus, at the known values of P, B_{cut} and H_{seam} in (14), the capacity of a mining machine (feed velocity) is unambiguously determined by the inverse proportion to the energy intensity of destruction (cutting).

Since the energy of cutting is spent to deformation of rocks in the cutting zone, as well as to separation and partial grinding of chips, the total energy intensity of the process is composed of the energy intensities of each process stage, is an integral characteristic of the process and, consequently, can be used to estimate breakability of coal.

On the other hand, under reference modes, the cutting energy intensity is directly proportional to the cutting force Z and inversely proportional to the cut cross-section S governed by the brittle–plastic properties of coal, which are commonly assessed by the side cutting edge angle φ:

$$H_W = k\frac{Z}{S} = \frac{ZL}{V} = \frac{A}{(b + h \cdot tn(\varphi))} \tag{15}$$

where k is the constant of proportionality; L is the cutting path, cm; V is the volume of broken material, cm^3; A is the cuttability, kgf/cm; b is the width of the active part of cutter, cm; h is the cut depth, cm.

The research into cutting of cement-and-coal blocks experimentally found the interaction:

$$tn(\varphi) = Bh^{-0.5} \tag{16}$$

Placing the latter relation in (15) yields:

$$H_W = \frac{A}{(b + Bh^{-0.5})} \tag{17}$$

In Russian mining practice B is named as the index of brittleness and describes the influence exerted by brittle–plastic properties of coal on energy intensity of cutting. The higher B means that coal is more brittle and its cutting consumes less energy, the other conditions being equal.

Thus, brittleness is understood as ability of coal to break with a certain energy at the same cuttability and constant parameters of the cutting mode. The range of brittleness is: B = 1.3…8.0. Qualitatively, by the index B, coal is split into viscous (B<2.1), brittle (B>2.1…3.5) and very brittle (B>3.5).

It is also found, that the value of B correlates with the index m denoted as the coal grindability:

$$B = \frac{e^{2.3m}}{m^2 - 8.4} \quad (18)$$

Numerical values of m are determined from the nomograms plotted based on the data of screen analysis of coal sampled from face area. To this effect, the averaged undersize values are plotted as functions of mesh sizes on the functional network of distribution of granulometric constituents (figure 2). Then, the resultant points are used to draw a straight line, and the index m is determined as the tangent of the angle to the X-axis: $m = b/a$.

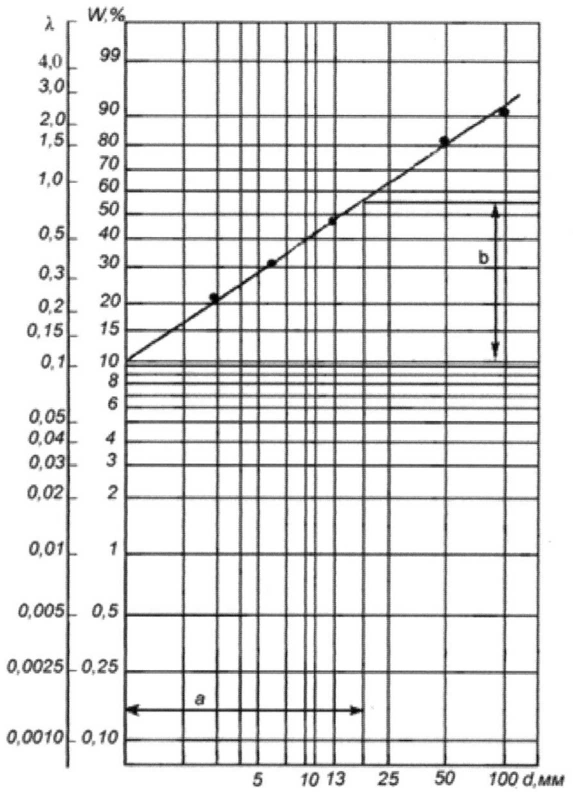

Figure 2. Graphical method of coal grindability determination.

It is found, that the grains-size distribution of broken coal obeys the law:

$$W = 1 - \exp(\lambda d^{m}) \quad (19)$$

where W is the total output (fractions of sample mass) of broken coal penetrated through the sieve with the mesh size d, mm; λ is the grindability parameter governed by the accepted methods and modes of cutting.

The index of coal grindability is taken as a basis in engineering design of coal brittleness and is used in classification of coal seams by dust-making ability (Bilgin, Copur and Balci, 2014).

Considering that cutting energy intensity is governed by the cuttability A and brittleness B in the formula (17), it is proposed to rank coal seams into seven breakability categories—from very weak to extra strong. In this case, breakability of coal and coal seams under mechanical attacks is interpreted as the ability to break at the certain energy consumption under constant operating conditions. Numerically, the breakability criterion R (kW·h·cm/m^3) of coal seam is found from the expression:

$$R = 0.38\frac{A}{B+1} \tag{20}$$

where A is the cuttability defined by the coal cuttability A_{coal}, or by the integral cuttability of coal seam A_e.

It is found that R varies from 2.4 to 65 in separate coal seams, which is a wide range sufficient for solution of engineering problems. It is analytically determined that the higher influence on R is exerted by the index of cuttability: as it grows from 60 to 480 N/mm at the constant B, the value of breakability increases 7–14 times. Oppositely, under the variation in brittleness at the constant cuttability, R raises only 1.6–3.2 times.

Figure 3 shows the nomogram for determination of breakability of coal seams. Here, the boundaries of each category are the lines of equal energy consumption.

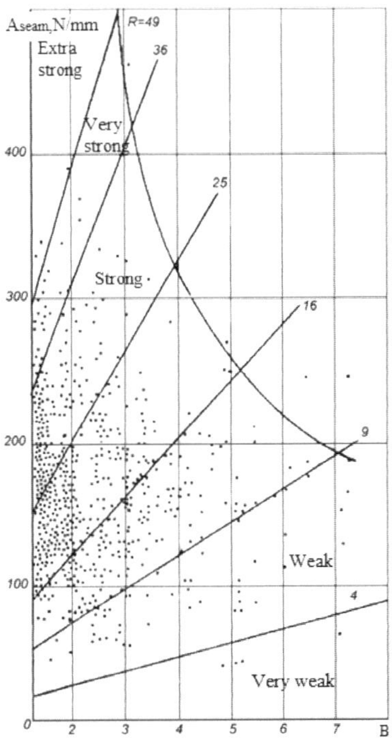

Figure 3. Nomogram to determine values and categories of coal breakability.

The points in Figure 3 stand for coal seams with different values of A_{seam} and B. The curve depicts the probable boundary of change in the values of B.

Abrasiveness of coal seams is governed by dirt bands and hard inclusions present in the seam. Estimation of coal abrasiveness uses the abrasiveness index ρ (mg/km) found experimentally. The procedure of coal seam abrasiveness estimation (Evans I, 1962) is based on attrition of a reference tool made of steel grade 45 having hardness HRC 24 ± 1 in reference operating conditions. The attrition test specimen is made of crushed components (coal, hard inclusions, dirt bands) taken proportionally to their content in coal seam. During the attrition tests, the pressure at the specimen–tool contact is constant.

With the known data on abrasiveness of coal seam components, the overall abrasiveness can be calculated from the formula:

$$\rho_{seam} = \rho_{coal}S_{coal} = \sum_{i=1}^{n}\rho_{dirt_i}S_{dirt_i} + \sum_{i=1}^{m}\rho_{hard_i}S_{hard_i} \tag{21}$$

where $\rho_{coal}, \rho_{dirt_i}, \rho_{hard_i}$ are, respectively, the average abrasiveness of coal, dirt bands and inclusions of an *i*-th lithology type; *n* and *m* are the numbers of lithology types of dirt bands and hard inclusions; $S_{coal}, S_{dirt_i}, S_{hard_i}$ are the densities (content) of the components in coal seam.

Systematization of data on breakability of coal seams and, owing to this, possibility to carry out an integrated assessment of strength characteristics of coal has made a framework for classification of coal seams as media destroyed by rock-breaking tools of mining machines (Dokukin A.V., Chirkov S.E and Norel B.K, 1981). On the whole, the classification is based on typification of coal seam structure with regard to different lithology types of dirt bands and various-size hard inclusions in the seams, as well as on division of coal seams into categories of breakability as function of cuttability and brittle–plastic properties of coal. Typification of coal seams by geological structure is given in table 3.

From the analysis of Table 3 data, as cuttability and density of inhomogeneities in coal seam increases, the generalized index naturally grows. For instance, in simple structure seams, or in seams containing small dirt bands and hard inclusions (typical condition group 1), the generalized index ranges between 0 and 52 N/mm, while in group 3 (high density and hardness of inhomogeneities), the index jumps to 523 N/mm and is on average 163 N/mm. Practical experience shows that such conditions limit productivity and reliability of coal mining machines.

Table 3. Density of hard inclusions in coal seam

Group of structural complexity of coal seams	Description	Generalized index A^*_{inh}
1	Seams of pure coal or with bands of coaly argillite and (or) argillite with cuttability A_{dirt} < 200 N/mm, or (and) with fractured hard inclusions with density to 1%	$\frac{0-52}{2.0}$
2	Seams with bands of argillite or (and) coaly argillite, or (and) siltstone with 200 < $A_{dirt} \le$ 400 N/mm, or (and) with fractured hard inclusions with density more than 1%, or with unfractured and consolidated inclusions with density to 2.5%	$\frac{1-163}{34.4}$
3	Seams with bands of siltstone, limestone and sandstone with A_{dirt} > 400 N/mm, or (and) unfractured hard inclusions with density more than 2.5% and consolidated inclusions of any density	$\frac{4-523}{163}$

If a coal seam encloses dirt bands and inclusions, it should be placed in a group that describes more complex properties of one of the components. Typification of coal seams by the breakability category is given in table 4.

Table 4. Typification of coal seams by breakability

Breakability category	Coal cuttability A_{coal} and brittleness *B*
I	Viscous coal (*B* < 2.1) with A_{coal} < 150, or brittle coal with A_{coal} < 200
II	Viscous coal with A_{coal} = 151–240, or brittle coal with A_{coal} = 201–300
III	Viscous coal with A_{coal} > 240, or brittle coal with A_{coal} > 300

These typifications are the convenient tool for sound engineering decision-making on selection of coal mining equipment (in particular, cutters or rock-breaking tools), or for solving some technological tasks. In this case, the typification can be compiled into a classification matrix (table 5) where each window describes a certain group of structural complexity and category of breakability, and can be assumed as a cumulative classification index.

Table 5. Classification matrix of coal seam breakability

Group of structural complexity of coal seams	Breakability category		
	I	II	III
1	1–I	1–II	1–III
2	2–I	2–II	2–III
3	3–I	3–II	3–III

4. Discussion

To prove the validity of the proposed selection method of integral evaluation of breakability of coal seams we analyzed the dependencies of the overhaul life of screw cutterheads R_{tbo} (Time Between Operations) on the following characteristics of coal seams: coal cuttability A_{coal}, which determines the overall load of cutters; generalized index of the contents and properties of the inhomogeneities in the seam A^*_{inh} , which is a given characteristic of the resistance to cutting of rock layers and solid inclusions and describing the characteristics of dynamism (uneven) process of destruction of the coal seam; specific heterogeneity content in seam $S^*_{inh} = S^*_{hard} + S^*_{dirt}$ reflecting, ceteris paribus, the frequency of peak loads that are responsible for breakdown failures of screw elements; abrasiveness of coal seam ρ_{seam} affecting screw failure by wear; integral cuttability of coal seam A_e, which takes into account the indicators A_{coal}, A^*_{inh} , and S^*_{inh} .

Figure 4 represents the results of the analysis of the set of values for overhaul life R_{tbo} of screw cutterhead with a diameter and working width of 800 mm equipped with picks. It shows that there is an explicit correlation of values for dependency $R_{tbo} = f(A_e)$ and also a lighter correlation in the dependency $R_{tbo} = f(A_{coal})$. However, there is an unacceptably high scatter for small values of A_{coal} and R_{tbo}. Table 6 represents the results of statistical checking of significance for correlations between R_{tbo} and A_e and between R_{tbo} and A_{coal}.

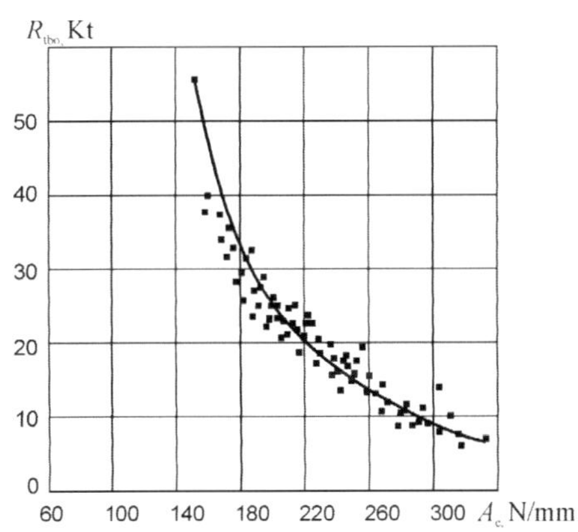

a)

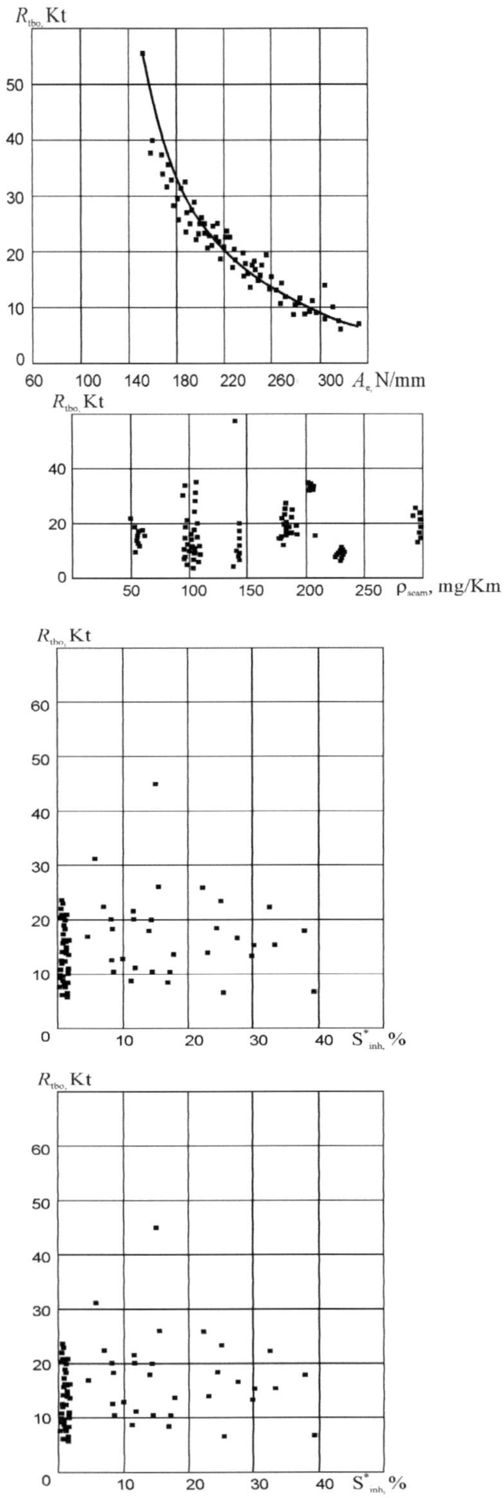

b)

c)

d)

e)

Figure 4. Dependencies of the overhaul life of screw cutterheads R_{tbo} on coal cuttability A_{coal} (a), integral cuttability of coal seam A_e (b), abrasiveness of coal seam ρ_{seam} (c), generalized index of the contents and properties of the inhomogeneities in the seam A^{*}_{inh} (d), and specific heterogeneity content in seam S^{*}_{inh} (e)

Figure 4 and table 6 prove the validity of the choice of A_e as a characteristic for the aggregate assessment of the cuttability of coal seams as objects destroyed by the cutters of the cutterheads of coal-mining machines.

Table 6. Results of statistical checking of significance for correlations between R_{tbo} and A_e and between R_{tbo} and A_{coal}

Dependency	Correlation index	Student's criterion	Fisher's criterion
$R_{tbo} = f(A_{coal})$	0.71	14.1	2.0
$R_{tbo} = f(A_e)$	0.88	32.6	3.9

5. Conclusion

In the presented study, we conducted an analysis of indicators that characterize the property of coal seams. In the work, strength assessment was carried out using the integral cutting ability, which was presented as the sum of the cutting ability, density index and heterogeneity of the medium. The study analyzed the Russian experience in assessing the hardness of formations and proposed an empirical expression to evaluate the strength and hardness of coal seams.

References

[1] Gibadullin A A et al 2019 *IOP Conf. Ser.: Mater. Sci. Eng.* **537** 042065
[2] Morkovkin D E et al 2019 *J. Phys.: Conf. Ser.* **1399** 033042
[3] Gibadullin A A, Kerimova Ch V, Chorshanbiev S R, Shmanev S V, Romanova Ju A and Sozontov A A 2020 *IOP Conf. Ser.: Mater. Sci. Eng.* **734** 012170
[4] Rakuta N V and Sitokhova T E 2019 Conceptual approaches to the formation of regional economic clusters *Upravlenie* **7(4)** 76-83
[5] Alber M, Yarali O, Dahl F, Bruland A, Kaesling H, Michalakopoulos T N and Cardu M. 2014 *Rock Mech. and Rock Eng.* **1(47)** 261-6
[6] Beron A, Kazanskii A, Leibov B and Pozin E 1962 *Coal Cutting* (Moscow: Skochinskiy mining institute, USSR) 234
[7] Bilgin N 1977 *Investigations into the mechanical cutting characteristics of some medium and high strength rocks* (Ph.D. Dissertation, University of Newcastle Upon Tyne) 562
[8] Bilgin N, Copur H and Balci C 2014 *Mechanical Excavation in Mining and Civil Industries* (Boca Raton: Taylor & Francis/CRC) 218
[9] Dokukin A V, Chirkov S E and Norel B K 1981 *Modeling of the Ultimate Stress State of Coal Seams* (Moscow: Nauka, USSR) 325
[10] Evans I 1962 A theory of the basic mechanics of coal ploughing *Proceedings of the International Symposium on Mining Research* **2** 761-8
[11] Evans I and Pomeroy C 1966 *The Strength, Fracture and Workability of Coal* (Oxford: Pergamon) 416
[12] Goktan R M 1997 A suggested improvement on Evans' cutting theory for conical bits *Proceedings of Fourth Symposium on Mine Mechanization Automation* **2** 96-101
[13] Hoyoung J and Seokwon J 2018 *Geomech. and Eng. An Int'l J.* **15(3)** 811-22
[14] Inyang H 2002 *Int. J. of Surf. Mining, Reclamation and Env.* **16(4)** 248-60
[15] Jienan P, Haichao W, Kai W and Qinghe N 2014 *Geomech. and Eng. An Int'l J.* **6(6)** 613-24
[16] Li X 2013 *The Sci. World J.* **5** 62-74
[17] Peijian J, Enyuan W and Dazhao S 2017 *Geomech. and Eng. An Int'l J.* **12(4)** 627-37
[18] Rojek J, Oñate E and Labra C 2011 *Int. J. of Rock Mech. and Mining Sci.* **48(6)** 996-1010

[19] Su O and Akcin N A 2011 *Int. J. of Rock Mech. and Mining Sci.* **48(3)** 434-42

[20] Pozin E Z, Melamed V Z and Ton V V 1984 *Coal Destruction by Mining Machines* (Moscow, Nedra) 246

[21] Leibov B M 1962 *Method of determination of coals cuttability results of mechanical tests of samples of any form* (Moscow: Skochinskiy mining institute) 146

[22] Linnik Yu N, Merzlyakov V G and Linnik V Yu 2010 Typification of coal seams with a view to selecting heading and mining machines and evaluating their availability for efficient operation *Marksheider Vestnik* **5** 12-8

[23] Pozin E Z 1964 *Instructions for determining the resistance of coal destruction using dynamometer drill SDM-1* (Moscow: Skochinskiy mining institute) 378

[24] Kuntysh M F and Baronskaya E I 1980 *Methods for estimating the properties of coal seams of complex structure* (Moscow: Nauka) 329

[25] Pozin E Z, Melamed V Z and Azovceva S M 1977 *Crushing of coal during cutting* (Moscow: Nauka) 415

[26] Pozin E Z, Melamed V Z and Azovtseva S M 1974 Determination Procedure of Coal Grindability in Cutting (Moscow: Skochinskiy mining institute) 187

[27] Zhabin A B, Polyakov AV, Averin E A and Sarychev VI 2018 State of scientific researches in the field of rock destruction by picks at the turn of the century *Proceedings of the Tula States University-sciences of Earth* **1** 230-47

IOP Conf. Series: Materials Science and Engineering **837** (2020) 012010 doi:10.1088/1757-899X/837/1/012010

Digital technologies as a tool for improving production and technological activities and information security in Fuel and energy complex

E I Dzhafarov and N A Kharitonova

Financial University under the Government of the Russian Federation, 49, Leningradsky Prospekt, Moscow, 125993, Russia

E-mail: elmircloud@gmail.com

Abstract. The article considers the features of using digital technologies in organizations of the fuel and energy complex (FEC). The characteristic features and main problems in the production and auxiliary processes of the oil and gas industry are identified. A study and analysis of the use of blockchain technology based on a variety of distributed registry in the fuel and energy sector, in particular for the oil and gas industry. The encryption algorithm used in blockchain technology is considered. The most relevant models for the application of projects based on distributed database subtechnology in production, trade in oil and natural gas, document management, logistics, with an assessment of the effect of digital technologies in the degree of influence on information security have been developed.

1. Introduction

A feature of the world economic system in the first quarter of the XXI century is the widespread digitalization of production and management processes in all areas of social production, including the extraction and processing of primary energy resources. The development of innovative information technologies contributes to the transformation of traditional methods of processing large amounts of information, which increases the efficiency of business-processes, allow you to form a competitive environment in all corporate areas, while ensuring the stability of the system itself, as well as reducing transaction costs, along with increasing the speed of business operations. The use of blockchain technology, based on the use of distributed registry methodology, allows implementing a reliable and transparent data management system for all participants in production and management processes, and is currently quite widespread around the world [1].

Despite the fact that initially modern information technologies developed in the banking sector, oil and gas companies actively use them in the exploration of oil and gas fields, conducting engineering and survey work , drilling wells, organizing logistics processes, making payments to consumers. At the same time, the reliability of the created information systems, as well as the efficiency of their functioning, is of particular importance.

2. Materials and methods

The authors, based on the study and processing of publicly available information, formed an idea of the degree of penetration of information and digital technologies in the production and economic activities of oil and gas companies, whose distinctive features are the large volume of property assets used, the importance of controlling the production, distribution and transportation of primary energy resources, and the high cost of investment IT projects. It should also be emphasized that these companies are characterized by specific problems (high operating costs, a constantly decreasing percentage of oil production at existing wells and the need to develop new fields not only to increase the production of hydrocarbon resources, but also to maintain previously achieved positions, as well as the instability of global industry markets and a fairly high tax burden on business).

In order to overcome the existing problems, the largest Russian oil and gas market players are developing not only in the direction of improving production and technological activities, but also introducing methods to improve the efficiency of energy resources production through the development of the following information and digital technologies [2]:

- artificial intelligence-used for managing logistics systems;
- algorithms for processing unstructured data (Big Data) in order to provide decision support in management activities;
- augmented reality is a tool for remote consulting , which allows employees working in remote fields to interact with experts, and managers to make more effective management decisions;
- digitalization of the field allows using modern technologies to virtually simulate all stages of field development, which, of course, reduces management risks.

Nedeco oil and gas company is the founder of the Bank, which provides it with comprehensive services using modern efficient banking decisions, such as [3]:

- mobile applications for remote cash flow management (online payment services, online transfer services);
- cloud sales registers and smart terminals, as well as mass payment services.

An example of such cooperation is the partnership of the all-Russian regional development Bank and Rosneft.

3. Results

The main products of financial technologies that are becoming widespread in the Russian oil and gas industry are.

Blockchain technology, which is a continuous chain of data blocks, where each subsequent block is linked to the previous one through a set of records, which allows you to perform any operation with contractors and partners faster, cheaper and more reliable.

The principle of blockchain as a sub-technology of distributed registries is to add information to each block containing data about operations with a time stamp, which is designated as a hash sum, i.e. values in hexadecimal form, which is also a digital signature. The hash function is necessary for compressing the original generated text of information into a short block of data. Then these blocks are built into a sequential chain of blocks. Visually, the block chain can be represented as follows figure 1.

A large array of data in the process of interaction between functional elements determines the optimal conditions for using blockchain technology. Based on the block chain, you can develop a single network for digitizing all major processes and automating them.

The oil-producing ecosystem is quite extensive, complex, geographically dispersed, but at the same time it is characterized by stable relationships of its participants and a large number of tasks in the field of production management.

Artificial intelligence that allows you to perform many operations in oil companies more efficiently and, most importantly, reliably. These are, first, smart wells and smart fields, which are innovative developments based on digitalization of production processes that can independently adapt to changing conditions. In addition, "smart well" or "smart field" is able to diagnose the system, which allows timely repair of technological equipment and ensure a more complete utilization of existing production capacity [5].

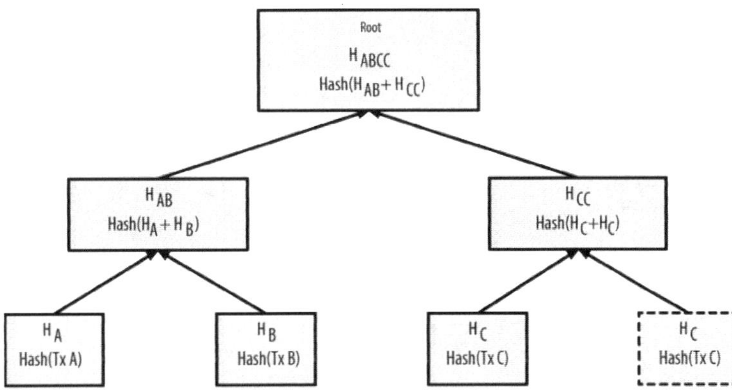

Figure 1. The hierarchy of blocks in the blockchain mechanism.

Smart – contract: an algorithm designed for entering into and executing contracts based on the fulfillment of all the conditions specified in it, which reduces the risk of errors and increases the speed of business transactions. The oil and gas sector is characterized by a large number of contractors, which implies the conclusion of a large number of contracts. When a person controls the execution of contracts, there is a possibility of technical errors and even abuse. Smart contracts, being self-executing, allow you to exchange assets without resorting to services, although they can be changed with the consent of the parties in respect of all amendments to the contract.

Big Data – technology that provides storage, processing and use of large volumes of information, which is especially important for oil companies that are engaged not only in production, but also in transportation to consumers of energy resources of the required quality.

Non-banking services that provide remote access to digital services.

As the main advantages of the ubiquitous spread of digital technologies, it is customary to highlight:

- reducing the risk of fraud and money laundering;
- improving the efficiency of companies by reducing transaction costs.

In General, the wider use of digital technologies should contribute to achieving the goals of the «Energy strategy of the Russian Federation until 2030», so as:

- increase the oil recovery rate by 5-10% in «digital fields»;
- reduce operating costs at digital fields by 10%;
- reducing capital expenditures on «digital fields» by up to 15% [6].

At the same time, digitalization of all components of the main production and management processes is impossible without improving information security systems, since in recent years the number of hacker attacks has been growing in direct proportion to the development of information technologies, which leads to significant losses and losses for oil and gas companies. According to estimates of Russian law enforcement agencies, the damage from the negative impact on the information resources of companies engaged in oil and gas production may amount to about 1.2% of GDP. The attacks are mainly aimed at extracting information about personal data and information about customers and transactions as in figure 2 [7].

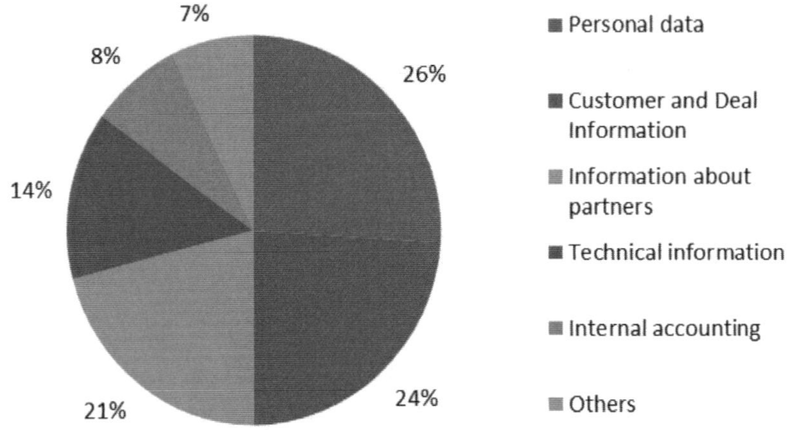

Figure 2. Incidents of information loss in oil and gas companies.

Unauthorized theft of information, expressed in the form of espionage, is usually aimed at obtaining technical information containing information about the features of the organization of technological processes, as well as production accounting data in order to actually determine the capabilities of companies to participate in certain investment projects.

Considering the side of information security, the block hierarchy described in figure 1 is also provided with encryption, that is, it is supplemented with cryptographic protection. In order to determine how the blockchain technology contributes to the protection of information and prevents its theft, let's turn to the simplified scheme for implementing cryptographic protection, shown in the figure 3.

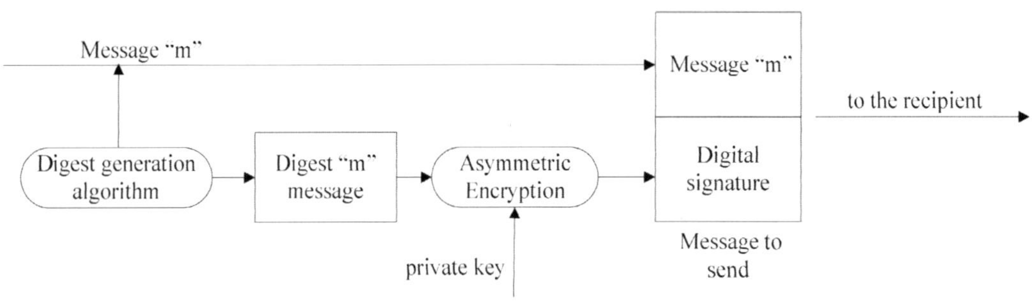

Figure 3. Scheme implementation of cryptographic protection.

When initiating an operation, two key types are generated in the distributed registry system: a secret key and a public key. The public key is determined from the paired secret key and passed to the rest of the network participants for verification of the electronic signature. A digital signature, as mentioned earlier, is a hash function of generated information. Then the initiator of sending information encrypts the information sent by the hash block with a secret key. The generated number pair is a digital signature. The result of implementing the blockchain encryption algorithm is to transmit a message to the recipient along with a digital signature.

Thus, the described encryption algorithm applied to each generated block in the chain defines the blockchain technology as a fairly effective tool to prevent hacking by cyber attackers. With the expansion of the use of digital technologies, the need to use blockchain technology is steadily growing in parallel with the growth of cyber threats, which requires timely and rapid implementation of

IOP Conf. Series: Materials Science and Engineering **837** (2020) 012010 doi:10.1088/1757-899X/837/1/012010

measures to prevent them or minimize the possible consequences associated with the «leak» of information.

4. Discussion

As noted earlier, the widespread digitalization of production and management processes that accompany oil and gas production has made it possible to switch to new technologies, despite the previously skeptical attitude to the idea of distributed registries. In particular, major international oil traders Mercuria and Trafigura have distinguished themselves by using blockchain projects to optimize business processes in resource trading, increasing the efficiency of transactions in hydrocarbon and gaseous raw materials and ensuring digital control of document flow. In an attempt to compete with existing crypto currencies, the company R Fintech has created its own virtual currency Bilur, the value of which is tied to oil assets. Prospects for the use of blockchain technology determine limitless opportunities, forming a modification vector for optimizing and reengineering ongoing processes in vertically integrated oil and gas companies, followed by the creation of new business models based on technologies built on continuous sequential chains of blocks.

It should be noted another advantage of the use of technology distributed register – cyber security. Cyber risks are a set of risks associated with a destructive impact on information systems and networks through the implantation of malware, which, in turn, have a negative impact on both infrastructure technologies and the information security of the company as a whole [8].

In recent years, the number of hacker attacks has been steadily increasing, which leads to significant losses and losses, the value of which, according to Russian law enforcement agencies, can be about 1.2% of GDP. At the same time, experts identified the main channels of cyber attacks on fuel and energy facilities through malware table 1.

Table 1. Characteristics of studied emulsions.

Types of malware	Frequency of implementation
Attacks on fuel and energy complex	40.2
Attacks on organizations that produce equipment and components for the fuel and energy complex	36.4
Ransom ware	11.7
Miscellaneous	11.7

The wider spread of distributed registry technology automatically ensures the safety of data on decentralized network nodes, while adding cryptography to the blocks of information, which ensures a high and reliable level of protection of commercial information.

The wider spread of distributed registry technology automatically ensures the safety of data on decentralized network nodes, while adding cryptography to the blocks of information, which ensures a high and reliable level of protection of commercial information.

5. Conclusion

Every business process in oil and gas companies requires infrastructure technologies that allow controlling the accuracy and reliability of the characteristics of the extracted and sold raw materials, facilitating and solving problems of speed of operations, as well as automating most of the procedures related to confirming the reliability of the results of production and economic activities and ensuring the guarantees of contracts. At the same time, the introduction of technologies that increase the efficiency of processes in companies must be accompanied by information security, since the development of digital technologies increases the risk of unauthorized interference and information leakage, which can threaten not only the reputation, but also, often, the very existence of the company.

References

[1] Brom A E and Terenteva Z S 2018 The use of blockchain technology in the lifecycle

management of products *Bulletin of the Volga state University named after V N Tatishchev* **2** 1 118-24
[2] Morkovkin D E et al 2019 *J. Phys.: Conf. Ser.* **1399** 033042
[3] Ivanova I A et al 2020 *IOP Conf. Ser.: Earth Environ. Sci.* **421** 032039
[4] Ermakov I A and Kuzminykh S S 2019 Application of distributed registry technology as one of the mechanisms for digital integration of supply chains *E-Managemen* **2** 45-58
[5] Sargsyan K R 2018 Development and prospects of using blockchain technologies in the global oil and gas industry *Innovation and investment* **8** 116-20
[6] *Russian statistical yearbook* 2018 (Moscow: Rosstat) 694
[7] *Ministry of Energy of the Russian Federation* Retrieved from: https://minenergo.gov.ru
[8] Pertseva C 2018 Fintech industry and information security *World and national economy* **4** 5-12
[9] Zimnukhova D I et al 2019 *J. Phys.: Conf. Ser.* **1399** 033097
[10] Pulyaeva V N et al 2019 *IOP Conf. Ser.: Mater. Sci. Eng.* **537** 042033

IOP Conf. Series: Materials Science and Engineering **837** (2020) 012011 doi:10.1088/1757-899X/837/1/012011

Differentiation of power supply systems for consumers in the Arctic zone

M M Gubanov[1], D E Morkovkin[1], Eraj Begmurodzoda[2], F F Sharipov[3] and Ju A Romanova[4]

[1] Financial University under the Government of the Russian Federation, 49, Leningradsky avenue, Moscow, 125993, Russian Federation
[2] Tajik National University, 2, Lohuti street, Dushanbe city, 734025, Republic of Tajikistan
[3] State University of Management, 99, Ryazan Avenue, Moscow, 109542, Russian Federation
[4] Moscow state University of technology and management K.G.Razumovsky (PKU), 73, Zemlyanoy Val street, Moscow, 109004, Russian Federation

E-mail: morkovkinde@mail.ru

Abstract. The article discusses the problems of providing energy supply to consumers in remote regions of Russia, including in the Arctic zone. The necessity of introducing and developing the best available technologies for autonomous energy supply to consumers in the decentralized electricity supply zone, which should be based on the combined use of technologies and reducing cross-subsidization for the long term in the framework of the implementation of a unified state energy policy, is proved. Differentiation of power supply systems of consumers was carried out, according to the results of which it is advisable to pay special attention to the location of the source of electric energy generation. The paper suggests that in the conditions of the Arctic zone it should be as close as possible to the center of energy consumption and the possibility of combined generation of electric and thermal energy.

1. Introduction

In order to solve the problems of ensuring energy supply to all consumers in the Arctic, it initially seems expedient to single out two main areas: the development of centralized power supply systems for consumers and the development of decentralized power supply systems for consumers. It is an obvious fact that ensuring reliable energy supply of the underdeveloped extended territories of the Arctic zone remote from the existing energy infrastructure in severe climatic conditions only through the creation of large power plants and the development of electric networks is technically and economically impractical, while the availability of energy sources for autonomous small consumers is the basis for the livelihoods of the population and to prevent depopulation of the country's territories [1-4].

The main priorities in ensuring energy supply to consumers in the Arctic zone are enshrined in the strategy and include [5-8]:

- Construction of nuclear thermal power plants;

IOP Conf. Series: Materials Science and Engineering **837** (2020) 012011 doi:10.1088/1757-899X/837/1/012011

- Optimization of the "northern delivery" of fuel through the use of renewable and alternative energy sources, as well as the modernization and reconstruction of energy plants with a developed resource and the introduction of energy-saving technologies and materials;
- Increasing energy efficiency and expanding the use of renewable energy sources (hereinafter - RES);
- Modernization and updating of housing and communal services using modern energy-saving technologies;
- Creation and implementation of economic mechanisms for the rational use of raw and biological resources, resource conservation and utilization of associated petroleum gas in areas of oil production.

2. Materials and methods
The purpose of this study is to differentiate consumer power supply systems by the criterion of the location of the source of electricity generation in the Arctic zone. To achieve this goal, the following tasks:

- A comprehensive analysis of the problems of providing energy supply to consumers of electric energy in remote regions of Russia;
- Verification of the hypothesis of the introduction and development of the best available technologies for autonomous energy supply to consumers of electric energy in the zone of decentralized electricity supply.

In the process of work, general scientific and specific methods of system analysis of the object, the method of analogy and comparison, the method of expert evaluations, which allowed the authors to solve the tasks, were used.

3. Results
In accordance with the above priorities, it is advisable to identify the following ways of developing centralized power supply systems: renovation of the electric power complex with the replacement of obsolete and physically obsolete equipment, increasing the capacity of intersystem sections to ensure the delivery of locked capacities, ensuring the technological connection of new consumers to existing ones and the creation of new power centers, increasing reliability of the power grid complex, rational integration in the energy system in new generation capacity through the use of nuclear power and renewable energy sources in order to minimize the fuel component in the final price for electricity, increasing electric power system management efficiency, reduction of losses in electric networks [9].

At the same time, the accession of new consumers to the Unified Energy System of Russia or a technologically isolated territorial electric energy system should be technically and economically justified during the development of a feasibility study with mandatory consideration of decentralized electricity supply options. The main criterion for making a decision should be its technical rationality and economic feasibility in accordance with long-term state goals, as well as the possible attraction of private capital. If the option of decentralized power supply is preferable, it is necessary to provide support for its implementation by presenting the best available technologies for autonomous power supply and state support for their implementation.

The integration of generating sources of electricity into the centralized power supply system and the point of their connection should also be justified as part of the development of a feasibility study, while in order to exclude additional costs for the power distribution scheme, the main priority should be the equivalence of replacement of inefficient or outdated generating capacities and minimal impact on prevailing power system structure. In the framework of this work, not only the full installed capacity of the power plant, but also the power of individual power units should be justified. If the emergence of a new generating source requires a change in the structure of the power system or the appearance of additional costs for providing a reserve of electric power and energy, they need to be taken into account in the integrated financial and economic model of the project [10-11].

It is unacceptable to take government decisions to provide external energy infrastructure to prospective consumers or to build new generating capacities in the Arctic zone with accession to existing energy systems without a deep study of this issue at the pre-project stage. The obtained results also need to be further worked out and taken into account in the framework of developing program documents that are controlled by the Ministry of Energy of Russia, authorities of entities, network and generation companies and the system operator, and only after that make rational decisions on the implementation of large investment projects.

In order to ensure the implementation of the action plan, it is advisable to differentiate the development of modern centralized power supply schemes for the Arctic zone and the adoption of relevant government decisions, as follows [12-14]:

- The power supply to energy-intensive consumers, the connected capacity of which exceeds 50 MW and the related development of the electric grid infrastructure with a voltage class of 330 kV;
- The associated development of electric grid infrastructure with a voltage class of 220 kV and higher;
- Power supply to other consumers and construction of power plants with a capacity of less than 25 MW.

The development of these program documents should be based on the principles of ensuring reliable and high-quality centralized power supply to consumers, as well as the priorities of state policy for the development of the Arctic zone.

Taking into account the priorities of the state energy policy development necessary small generating sources of electric and thermal energy commensurate with consumers located in close proximity to consumers and the possibility of issuing electric power in a low-voltage network of centralized electricity and heat supply circuit. Such an approach will reduce the load on centralized electric and local heating networks and, due to the distribution of consumers and, accordingly, renewable energy sources across the territory of the subject, will reduce the uneven generation of electric energy and power throughout the energy system as a whole. Thus, the ratio of large and small consumers and energy sources will be maintained in the energy system, which will guarantee the energy security of the regions [15].

Changing the existing structure of the electric power complex of the Arctic zone towards creating the above-described distributed generation system based on renewable energy will require the government to solve a significant number of scientific, technical, legal and economic of resources.

For decentralized energy supply systems in accordance with the above state policy priorities, it is advisable to identify the following development paths: improving the energy efficiency of buildings and structures through the development of rational architectural and structural solutions, the use of modern building and insulation materials, compliance with advanced energy-saving construction technologies to eliminate heat energy losses and maximizing the use of natural light (realizing the concept of a "passive house"; providing constructed and reconstructed buildings with modern engineering systems, including ventilation with heat recovery, energy-saving artificial lighting, integrated automation and multi-tariff commercial metering to ensure the efficient use of electric and thermal energy (implementation of the "smart house" concept); optimization of the use of imported fuels with an increase in the efficiency of the use of energy; optimization of fuel delivery routes to reduce the transport component in the cost of its delivery; the search for modern science-based solutions to ensure the production of electric and thermal energy in the Arctic zone with the goal of maximum replacement of imported fuel [16-17].

4. Discussion

In these conditions, the most rational solution is the development of comprehensive government support measures for the introduction and development of the best affordable technologies for autonomous energy supply to consumers in the decentralized electricity supply zone, which should be based on the combined socio-economic effect of using these technologies and reducing cross-subsidization for the long term in the implementation framework unified state energy policy.

In order to further consider various categories of consumers and the corresponding energy supply technologies, we will differentiate and conditionally distinguish the following three main groups of consumers in the decentralized power supply zone:

- Enterprises for the production of hydrocarbons;
- Mining enterprises;
- Large seaports of the Northern Sea Route;
- Offshore oil and gas fields;
- Small settlements (towns, villages, villages and others);
- Military facilities.

Consistent with consumer groups, the existing and most promising types of energy supply sources, which can be considered to solve energy problems and meet the needs of the development of the economy of the Arctic zone, were also conditionally divided into three main groups:

- Power plants using fossil fuels (gas, oil, coal);
- Nuclear power plants of low power (AFMM), including floating ones;
- Hybrid complexes based on a combination of fossil fuels and renewable energy sources (solar, wind energy and small hydro), renewable energy sources in combination with high-capacity drives, fuel cells and other small generating sources.

5. Conclusion

Sources and power supply schemes of these consumers must be selected in accordance with the schedule of electric and heat load of consumers, the magnitude of electric and heat consumption, and the reliability categories of electric and heat supply, the possibility of rational management of demand for electric and thermal energy, as well as long-term goals of economic activity of consumers. Particular attention should be paid to the location of the source of electric energy generation, which in the Arctic zone should be as close as possible to the center of energy consumption and the possibility of combined generation of electric and thermal energy.

References

[1] Alpidovskaya M L, Gryaznova A G and Sokolov D P 2018 Regress Economy Vs Progress Economy: "Alternatives Of Senses" *Advances in Intelligent Systems and Computing* **622** 638-46

[2] Repnikova V M, Bykova O N, Skryabin O O, Morkovkin D E and Novak L V 2019 Strategic aspects of innovative development of entrepreneurial entities in modern conditions *International Journal of Engineering and Advanced Technology* **8(4)** 32-5

[3] Sorokin D E 2015 Economic theory, economic reality and economic policy *Economy of region* **1(1)** 17-29

[4] *Russian statistical yearbook* 2018 (Moscow: Rosstat) 694

[5] Shumaev V A, Morkovkin D E, Nikonorova A V, Nezamaikin, V N and Yurzinova I L 2018 Innovative aspects of agritourism project management *SERVE 2017, Financial and Economic Tools Used in the World Hospitality Industry - Proceedings of the 5th International Conference on Management and Technology in Knowledge, Service, Tourism and Hospitality* 241-8

[6] *Ministry of Energy of the Russian Federation* Retrieved from: https://minenergo.gov.ru
[7] Ivanova I A et al 2019 *J. Phys.: Conf. Ser.* **1399** 033038
[8] Lopatkin D S et al 2019 *J. Phys.: Conf. Ser.* **1399** 033061
[9] *Energy Strategy of Russia until 2035* Retrieved from: https://minenergo.gov.ru/node/1920
[10] Tolkachev S A et al 2020 *IOP Conf. Ser.: Earth Environ. Sci.* **421** 032041
[11] Dudin M N, Zasko V N, Dontsova O I and Osokina I V 2020 The energy politics of the european union and the possibility to implement it in post-soviet states *International Journal of Energy Economics and Policy* **10(2)** 409-16
[12] Rudaya I L 2019 Issues of strategic management of Russian innovative organizations' intellectual property *Upravlenie* **7** 1 50-9
[13] Sadriddinov M I, Mezina T V, Morkovkin D E, Romanova Ju A and Gibadullin A A 2020 Assessment of technological development and economic sustainability of domestic industry in modern conditions *IOP Conference Series: Materials Science and Engineering* **734** 012051
[14] Ustyuzhanina E V, Evsukov S G, Ustyuzhanin V L and Novikova E S 2019 The distribution of economic rent within global value chains in resource management *International Journal of Supply Chain Management* **8(6)** 1018-26
[15] Yuryeva A A *et al* 2019 *J. Phys.: Conf. Ser.* **1399** 033099
[16] Morkovkin D, Lopatkin D, Sadriddinov M, Shushunova T, Gibadullin A and Golikova O 2020 Assessment of innovation activity in the countries of the world *E3S Web of Conferences* **157** 04015
[17] Pulyaeva V N *et al* 2019 *IOP Conf. Ser.: Mater. Sci. Eng.* **537** 042033

AUTHOR INDEX

Institute of Physics
Dirac House, Temple Back
Bristol BS1 6BE UK

ISSN: 1757-8981
ISBN 978-1-7138-2693-4

CURRAN ASSOCIATES INC.
proceedings.com

9 781713 826934